The NEUROSCIENCE of FAIR PLAY

The NEUROSCIENCE of FAIR PLAY

WHY WE (USUALLY)

FOLLOW THE GOLDEN RULE

DONALD W. PFAFF, PH.D.

With a Foreword by **EDWARD O. WILSON**

DANA PRESS   *New York • Washington, D.C.*

Published by Dana Press
New York/Washington, D.C.

The Dana Foundation

745 Fifth Avenue, Suite 900

New York, NY 10151

900 15th Street NW

Washington, DC 20005

DANA is a federally registered trademark.

ISBN-13: 978-1-932594-27-0

Library of Congress Cataloging-in-Publication Data

Pfaff, Donald W., 1939–

The neuroscience of fair play : why we (usually) follow

the Golden rule / by Donald W. Pfaff.

p. cm.

ISBN-13: 978-1-932594-27-0

1. Neuropsychology—Moral and ethical aspects.

2. Ethics—Physiological aspects. 3. Golden rule.

I. Title.

QP360.P4625 2007

174.2'968—dc22

2007024380

Cover Design by Christopher Scardina / William Stilwell

Text Design by Kachergis Book Design

www.dana.org

CONTENTS

ACKNOWLEDGMENTS

Initially, I wrote this book for other scientists as a form of "intellectual play" that might help us to think about more complex patterns of behavior than we usually do. But then, the evidence and arguments seemed so interesting that I thought many smart folks who do not do brain research for a living might want to read the book as well. As a working scientist, I did not feel capable of presenting the material clearly enough for that purpose, and sought help. The brilliant conceptualizations and expressive powers of my editor at Dana Press, Jane Nevins, have made all the difference in changing the book from an Annual Reviews of Neuroscience type of compendium into a presentation that might interest those outside my field. Additionally, science writers Sandra Ackerman and Luba Vikhanski have worked patiently with me to expand some explanations and limit others in order to achieve a smooth narrative flow. I am grateful for all of this help. Susan Strider, at the Rockefeller University, is a graphic artist whose skill made the illustrations clear and attractive.

FOREWORD

The Neuroscience of Fair Play establishes more persuasively than any previous book the cause-and-effect linkages between biology, psychology, and the humanities. It is on the one hand a superb introduction to neuroscience in the service of moral philosophy; and on the other, in reverse, it shows how interest in the innate traits of human nature have stimulated and guided some of the key recent advances in neuroscience. Donald W. Pfaff, a leading researcher in this intermediate field, delivers a crystal-clear tour through the relevant technical intricacies of the science. The ideas that emerge are among the most important in their relevance to human affairs. They inform philosophy because they have so much to say about the mind, but they also have many practical implications for psychology and medicine.

Pfaff shows us that psychological phenomena can now be tracked to a substantial degree from genes through neuronal activity to brain circuitry and behavior. The result of this outwardly esoteric research is a major advance in human self-understanding. It is a step up from the intuition and magic of old and an objective, verifiable account of origin, process, and history.

The author provides a "theory of the Golden Rule" to explain in naturalistic terms the general occurrence of altruism—not just to offspring, siblings, and other pedigree relatives, but also to complete strangers. Neuroscience has helped explain this noble oddity of human nature, because the brain is an organ not merely divided into major parts, but divided against itself. Thus, the primal fear triggered by frightening or stressful stimuli, a response becoming well understood at the molecular and cellular level, is counterbalanced by an automatic shutdown of fear-inducing memories when altruistic action is appropriate. The individual "loses" himself. He identifies with an individual in need, for example, or toward whom he has been aroused to the level of violent aggression.

To solve this paradox, Pfaff utilizes the distinction between the two levels of explanation that are now standard in modern evolutionary biology. The brain, a supremely complex system of interacting nerve cells, hormones, transmitters, and growing cells, creates processes that variously reinforce or cancel one another out, according to context. To understand the way the brain works to create the conscious mind and all the subterranean forces driving it is the *how* explanation. It identifies the "proximate" causes of mental activity. But scientists also need to learn *why* the brain works in such and such a way and not some other. They search for "ultimate" causes by deducing the evolution of the brain action. Among the key questions to be answered are those posed in *The Neuroscience of Fair Play:* why are people at times selfish and aggressive, yet in other circumstances altruistic, sometimes to the point of self-sacrifice— and moreover to complete strangers.

The answer to the second question is likely natural selection among competing groups as opposed to individuals within each competing group in turn. This is the perception of Golden Rule Theory, and it is shared by evolutionary biologists who address the same problem, including the present writer. The roots go back to Charles Darwin's *The Descent of Man*. Multilevel selection theory, as it is called, can be stated roughly as follows: groups composed of people who are smarter and more altruistic tend to prevail over those who are less so. If the traits, with all the many molecular and cellular processes that yield them, have a genetically varying predisposition, the traits will strengthen by evolution across the population of all groups together. Put in the framework of present-day knowledge, the multilevel approach, encompassing the *how* and the *why*, is now bringing the major biological disciplines together in the quest to explain the oddly conflicted nature of the human mind.

EDWARD O. WILSON

Pellegrino University Research Professor, Emeritus, at Harvard University and Honorary Curator in Entomology at the Museum of Comparative Zoology

The NEUROSCIENCE of FAIR PLAY

1

SUBWAY STORY

Many of us begin the New Year in a quiet, not to say gingerly, manner. Wesley Autrey began the year 2007 by putting his life on the line—literally—for a stranger.

On January 2, 2007, as Autrey, a black construction worker, navy veteran, and fifty-year-old father of two, was waiting for the subway train that would take him to his construction job in New York City, a young white man standing near him crumpled to the floor in a seizure. Several people went to help the afflicted man—a kid, really, barely out of his teens, who as it happened had been on the way to his drama class at the New York Film Academy when he suddenly found himself in this real-life drama instead. Autrey, thinking quickly, got a pen from another rider and put it between the young man's jaws to keep him from biting his tongue. The seizure passed and several people helped the boy to get up. He was okay, he assured them. Well, not quite.

Back on his feet a little too soon, perhaps, the young man still seemed a bit weak or dizzy. In any case, he toppled off the subway platform onto the tracks. Autrey jumped down to pull him back to safety—and it was at this point that the situation abruptly got worse, because a subway train appeared, heading straight toward them.

What happened next took only a fraction of a second. Autrey, calculating that he had enough time to climb to safety alone but not if he dragged the boy with him, chose instead to throw himself down on top of the stranger, shielding him with his body. The train rushed over both of them with only inches to spare. Autrey later explained this heroic action by saying that years of construction work had made him familiar with tight spaces; he had guessed that he and the boy would have enough room to survive underneath the train. "I had to make a split decision," he said—not a split-*second* decision, but one that cut two ways. Without question, Autrey had saved the young man's life. But why, exactly, did he do that? What in the world was going on in Mr. Autrey's brain to allow him to perform this incredibly altruistic act?

This book lays out a possible answer. A long tradition in philosophy claims we do not have to refer to supernatural phenomena in order to consider why people act in a good or a bad way—and I intend to keep my explanation strictly within the bounds of the natural and the scientific. The whole focus in these pages is on the possibility that some rules of behavior are universally embedded in the human brain—that we are "wired for good behavior."

As the head of a large neuroscience laboratory at a medical research institution, with more than 40 years of work and over 700 scientific publications to my name, I've been asked many times while writing this book what put me onto this topic in the first place. I have been studying the neurobiological mechanisms of behavior for several decades and have had a lifelong interest in human social behavior in all its stunning variety. I am fascinated by why one person may behave one way and another person behaves differently. And what determines temperament? Certainly one aspect of temperament relates to whether we act in a kind and gentle way toward others or whether we are thoughtless or uncaring or even cruel.

For several years now, I have been reading far and wide in the literature of religions throughout the world, looking to answer just one question: "Can I find an ethical command that seems to be true of all religions, across continents and across centuries?" Well, I found one, and you'll recognize it instantly. You probably know it as the Golden Rule. Explaining how humans manage to behave according to this command (most of the time) is the challenge of this book.

Actually, it's amazing that a neuroscientist can now consider this type of question within his domain. During my career as a laboratory scientist, including the research my lab is conducting right now, I have studied the biology of the most basic kinds of behavior, and I have been careful to delimit the scope of the problems we attack so that we have a realistic chance of success. But once I found abundant evidence for a universal ethical principle, I was convinced there must be a biological reason for it.

My thinking ran like this: If this type of behavior takes place throughout human society, it must come from some trait in human biology—not something we tell one another to do, but a feature, or, in biological terms, a mechanism, that exists in our physical being. But what could it be? That is the mystery, and, since the mechanism may well take many lab hours to pin down precisely (though I predict you'll ultimately come to agree with me about how it must work), what we are working toward in this book is neither more nor less than a theory about the neuroscience of ethical behavior.

Why do I call this a "theory"? For scientists, "theory" means much more than "maybe" or a "hunch." A theory is a hypothesis based on solid evidence, to be tested and replicated; the evidence leads to the idea—no wobbles, no distortion, no gaping holes. We use "theory" to get our colleagues to spend thousands of grant dollars and thousands of hours in the lab testing the idea or facilitating their own work with it—and, it goes without saying, heaven help you if your theory jumped to a conclusion or in some other way didn't get it right. My using the word to describe the topic of this book means I *know* the trail of evidence leads to how we are hardwired to follow the Golden Rule. It is a trail other scientists can follow, and I want them to. The most famous theory (or infamous, depending on your point of view) is Darwin's, still called a "theory" after 150 years of testing. I'd like to have this one play out that way.

I believe that we are wired to behave in an ethical manner toward others, and they toward us. But with all the life-supporting functions that the brain handles from one millisecond to the next, only a few are likely to be capable of sparking an ethical response. These must be circuits crucial to our survival: the ones that are active whenever a situation may suddenly or significantly change our current status. The circuits we are interested in light up in moments of crisis: when a child runs out into the street in front of your car; when you are hurrying to get out of the rain and a stranger in front of you slips on the sidewalk; when you turn on the television and see an appeal for disaster relief in a part of the world you know only from a map.

Let me be clear. In company with most of my fellow neuroscientists, I believe strongly that the brain does not have a signaling circuit dedicated to ethics—or to truth or to moral philosophy, for that matter. What the brain has is a mechanism to make use of circuits that are already there, in order to disable the self-preference that is akin to our instinct for survival. Something must happen that allows a decision to forbear from harm or to give help. What can overrule the desire to put ourselves first? And what can trump our emotional objection to self-restraint or inconvenience or merely admitting a mistake?

From a scientific point of view, the simplest explanations of any phenomena, even complex phenomena, are best. The explanation I propose is very simple in the sense that it does not suppose all kinds of special, superhuman capacities for people to act altruistically or avoid hurting each other—no fancy nervous contraptions, no big deal. It does not even require the individual to learn anything in order to behave in an ethical fashion. I simply posit a *loss of information*. That is an opaque neuroscientific expression, but a basic one, and so wonderfully easy to explain that a person with ordinary common sense, or even a neuroscientist, can easily believe it.

But there's more going on here than a momentary loss of information; my theory brings in another brain circuit as well. This circuit, which supports—indeed, urges—caring, helpful behavior, got its start with hormonal actions that foster sexual love and parental love. Thus, treating other people right is guaranteed both by the avoidance of hurtful acts and by the performance of loving acts. Mr. Autrey had more than one brain system going for him when he behaved in such a heroic fashion.

If these brain systems impelling us toward good behavior are so powerful, why is there so much evil in the world? Let's admit right up front that in the world of animal behavior, aggression is absolutely necessary for finding food and mates and for defending territory, and it is subject to the control of various biological systems. In the case of human society, I'm forced by the ubiquity of individual acts of violence and organized acts of war to recognize the existence of hormonal and brain mechanisms that cause aggression. A certain degree of aggression appears necessary in order for people to try new things, to take initiative, to get ahead. However, it gets out of hand in the form of both individual violence and vicious warfare.

Ultimately, our ethical behaviors depend on a *balance* between the forces in the nervous system that support society and the forces toward aggression. This balance can be influenced by small changes in our genes and by environmental events, especially during two critical periods in our lives: infancy and adolescence.

My theory is not just for my fellow brain researchers, and it will not call on you to track down masses of scientific literature or to absorb every specific neurobiological point presented in these pages. In the words of the eminent pediatrician Benjamin Spock, who wrote bestselling books of advice for parents long before "parent" was a verb, "You know more than you think you do." The chances are that as I explain how, in the face of evils of every kind, humans manage to behave in an ethical fashion, you'll come across much that feels familiar to you.

Even so, we are launched on an entirely new journey, putting forward an account of complex human social behaviors that no one has ever tried before. To be sure, it will be controversial. Philosophers are wrangling right now in scholarly journals over whether an account such as this can ever be proved or disproved. Some religious devotees may object to a natural scientific explanation of matters they consider within the realm of the divine. But the reward for trying is the achievement of a new scientific paradigm—an enlightenment in its own way—that helps us understand why we behave in a good and generous way when we do.

So let's begin. First, I argue that I have identified a rule governing our ethical behavior that is universal among humankind and that should be treated as a natural product of our human brains. By the end, I hope you'll agree that we understand better, on the basis of solid scientific evidence and well-buttressed theory, why Wesley Autrey did the wonderful thing he did.

2

THE GOLDEN RULE

Associated with every religious system I have read about is a norm known as the Golden Rule. In essence, it requires that I do unto you as I would have you do unto me. This rule is so ingrained in our social behavior as to be intellectually invisible. As a result, we have rarely stopped to question where it came from.

Some experts would claim this principle is a product of evolution. Individuals who behave altruistically—that is, they aid others, even at some cost to themselves—help their group to survive better and ultimately to produce more offspring like themselves. The impulses toward this behavior have been passed down along with the rest of the genetic code for the human brain, and now appear not only in behavior but also in brain activity that we can detect and track. If this is so, we can understand why this rule and its many variations have survived in human ethical systems, philosophies, and religions.

What I have in mind is not religion per se, because not all statements of this rule are religious in their appearance. In fact, most of us first encounter the principle in the form of a loaded question: "How would you feel if Tommy or Ruthie put sand in your hair/called you a nerd/broke the arm of your action figure?" Nor is it a concept that holds significance only within a narrowly defined group. If we were discussing, let's say, the social customs observed only on the South Side of Chicago during weekday thunderstorms, this would be a very different book. Instead, I as a neuroscientist am trying to explain in elegant scientific fashion how a discoverable set of brain mechanisms can account for behaviors that follow the Golden Rule in every human society.

The diversity of human cultures today covers an enormous range. Even now, in an age of globalization and cultural merging, more than 6,000 languages can be heard on this planet. How can we embrace all these particularities and still claim to identify a principle that works among all human social groups?

Account for a kind of behavior that exists in every society? How can a neuroscientist dare attempt this? It is because complex behaviors do not always require the most complex explanations. Odd as this may sound, neuroscience is replete with illustrations. To take an example from my own research, we have been able to discover the mechanisms underlying the sociosexual behavior known as "sex drive," although we do not yet have a full scientific explanation of the simple act of walking. Recently, some leading cognitive neuroscientists have challenged us to begin facing the ethical questions inherent in neurobiology and, indeed, to try to understand the ethical responses themselves. A natural, universal principle seems a good place to start.

Ever since the Greek philosopher Plato, thinkers have been sure that ethical behavior would be subject to natural explanations. Plato himself wrote that a morally correct life is happier than an immoral one, for at least three reasons. First, an immoral person's desires would not be limited by his morality and thus would expand to a point where they could not be satisfied. Second, an immoral person cannot even compare the pleasures of reason with those of unlimited desires, because he lacks the reason. Third, indulging physical pleasures is ultimately less satisfying than the pleasures of reasoned intellect. In other words, a morally correct life fosters one's own well-being.

All too often, philosophers and social scientists seem to feel their work is threatened by natural scientific explanations of complicated human behaviors. Nevertheless, following Plato, a tradition among moral philosophers has admitted the possibility that behavior according to ethical principles could be explained as a natural phenomenon. For example, the seventeenth-century Dutch philosopher Baruch Spinoza linked moral theory to biological theory—that is, thinking about what "ought to be done" could not be separated from thinking about the natural world. For Spinoza, each human being is simply a finite part of Nature, maintaining by his behavior a balance with other humans and the rest of Nature. Thus, ethical behaviors are rigidly determined by the laws of Nature. Spinoza anticipated a modern psychological approach to ethics in that he linked moral instincts to what he called the "Active Emotions," *animositas* and *generositas*—respectively, rational self-love (*animositas* derives from the Latin *anima,* or soul) and rational benevolence. He also anticipated that some actions counted as altruistic actually are performed in the service of clear-sighted, long-term self-interest.

Meanwhile, in England, Thomas Hobbes wrote that although we have to deal with "the contrariety of the Naturall Faculties of the Mind, one to another, as also of one Passion to another," which makes it difficult "that any one man should be sufficiently disposed to all sorts of Civill duty," nevertheless "Education and Discipline" and "the faculty of solid Reasoning" can lead each person to behave in a normal, prosocial manner. ("Prosocial" is a word that I like as the opposite of "antisocial": examples of behavior that tend to foster the capacity of human beings to live and work together.) In our time, Alasdair MacIntyre, professor emeritus in the Department of Philosophy at Duke University, points out that day-to-day human life requires values such as truth telling, promise keeping, and a balanced fairness among individuals.

How universal is the Golden Rule?

Plato, Spinoza, and the others were on to something. In a survey of several millennia of civilization and many different religions, I have not yet found one that does not include the Golden Rule.

From Asian origins, examples include the following:

A disciple asked the Chinese Master "Is there one word which may serve as a rule of practice for all one's life?" Confucius answered "Is reciprocity not such a word? Do not to others what you do not want done to yourself—this is what the word means." (Confucius, *Analects*, 15.23)

Pity the misfortunes of others; rejoice in the well-being of others; help those who are in want; save men in danger; rejoice in the success of others; and sympathize with their reverses, even as though you were in their place. (Taoist writings, Tai-Shang-Kan-Ying Pien)

From orthodox Hindu philosophy:

Do not unto others what ye do not wish done to yourself; and wish for others too what ye desire and long for, for yourself—This is the whole of Dharma, heed it well. (Vedic Scriptures, *Mahabharata*)

From a Persian religion, centered around Zoroaster, the prophet:

That which is good for all and any one, for whomsoever—that is good for me. . . . What I hold good for self, I should for all. (Zoroastrian writings, Gatha, 43.1)

From the Old Testament of the Bible:

Thou shalt love God above all things, and thy neighbor as thyself. (Leviticus)

From Islam:

Noblest religion this—that thou shouldst like for others what thou likest for thyself; and what thou feelest painful for thyself, hold that as painful for all others too. (Hadis, Sayings of Muhammad)

And from the New Testament of the Bible, the Christian Golden Rule:

Whatsoever ye would that men should do unto you, do ye even so to them. (Matthew 7:12)

As for Buddhism, its teachings contain no direct statement that is equivalent to the Golden Rule, although they are entirely consistent with it. This is not the paradox it seems. In the Buddhist realization of a universal existence, the specific words for "I" and "you" do not make sense. However, the Buddhists teach that our human existence is one with the universal spirit Anatman. Since each of us is part of the same spiritual entity, harming another person would inevitably bring harm to oneself. The Dalai Lama, leader of Tibetan Buddhism, teaches the supreme importance of compassion and altruism:

> With a determination to accomplish
> The highest welfare for all sentient beings.
>
> (gLang-ri-thang-pa, 1054–1123,
> *Eight Stanzas for Training the Mind*)

Indirect statements amounting to the Golden Rule also appear among the Navajo people of North America.

The universality of the Golden Rule is apparent outside of religious contexts, as well. In one of his most famous utterances, the great eighteenth-century philosopher Immanuel Kant proclaims his categorical imperative: "Act only according to that maxim by which you can at the same time will that it should become a universal law"—by which he meant, "Each person should do only what she or he would allow others to do."

For Kant, this imperative is the basis of all morality. In fact, it represents a fundamental way in which human rationality finds expression, since it requires lawful, self-consistent judgment, one that stems from the person's own, internal system of values. The phrase "I must" dictates proper social behavior in a way that is unconditional, to be followed regardless of the desires at the moment. Thus, Kant's ethical philosophy is thoroughly grounded in a consideration of our human physiological and psychological makeup.

The universality of the Golden Rule is like the ubiquity of the primary means of communication among human beings: speech, which has developed, albeit with slight variations, among every human society known. In the late 1950s, the outspoken MIT linguist Noam Chomsky rocked the scientific world with his claim that the human brain is predisposed to recognize and generate the complex patterns of sound and meaning that constitute a language; similarly, our brains may be predisposed toward observing the extensive and nuanced codes of conduct that make up ethical behavior.

Ethics across species and eons

Societies that observe the Golden Rule are found even beyond the human species. The literature on animal behavior is rife with observations of what looks almost like kindness among nonhuman primates, the animals most closely related to us in both biological and evolutionary terms. Some of the most celebrated examples are reports by Frans de Waal, of Emory University, on the predilection of the bonobos (a subspecies of chimpanzee) for civilized social organization—social organizations with stable rules that protect against aggression.

More recently, Joan Silk, at the University of California at Los

Angeles, writing in *Science* magazine, describes prosocial behavior among female baboons. Silk says the females that stay in close contact and often groom one another, combing through each other's fur (with the added bonus of eating any insects found therein), tend to produce more babies and raise them more successfully than female loners in the same troop. Hans Kummer, of the University of Zurich, Switzerland, cites examples from the older literature on other nonhuman primates, ranging from food sharing, sexual fidelity, and respecting one another's social relationships to helping a relative who is under attack. Habits like these among nonhuman primates make it easier to accept the universality of Golden Rule–like principles among human societies.

Evidence that our brains were constructed along these lines long ago comes from mounds of prehistory. In their recent book, *Inside the Neolithic Mind,* South African scholars David Lewis-Williams and David Pearce discuss the wide range of social practices they believe were exhibited by Neolithic peoples. They argue from archaeological and anthropological evidence that neural patterns behind these social practices were already hardwired into the human brain by the time of the Late Stone Age, about 10,000 years ago. According to Lewis-Williams and Pearce, "The commonalities we highlight cannot be explained in any other way than by the functioning of the universal human nervous system." Thus, the brain would express in its circuitry the biological forces that help to shape human behavior in society. And it would account for the Golden Rule—or, in biological terms, "reciprocal altruism."

It is almost trite to say that individuals take risks to protect the group, but such is the case. In war, soldiers take risks to shield others from dangers such as hand grenades and go to extraordinary measures to rescue others—for example, wounded fellow soldiers—when almost certain to be injured or killed while doing so. People rush into burning buildings to save occupants and jump into water to pull out drowning victims whom they have never met before.

Robert Hinde, an internationally recognized ethologist at Cambridge University, has taken on the task of explaining a genetic basis for certain moral behaviors. According to him, the easiest cases deal with the behavior of people who are related to each other. Since natural selection, the "engine" of evolution, operates by means of individuals passing on their own best genes—that is, the ones most likely to help the individual survive and reproduce—prosocial behaviors toward relatives are the behaviors of choice. In those cases, behavior that might normally be described as altruistic actually could be considered "selfish" genetically, because in fostering the welfare of our kin we are helping to pass on some of our own genes to the next generation.

What about altruistic behaviors toward those with whom we are not related? One theoretical approach is simply to view the behavior as part of an "exchange." In long-term relationships among unrelated people, as long as all parties are satisfied with the perceived outcomes of their exchanges, whether the rewards come now or later, then mutually cooperative, prosocial behaviors can be understood.

But what about truly unselfish behavior? After all, people donate money to charities, spend precious leisure time and energy in blood drives and volunteer fire departments, and commit other acts that appear purely unselfish. A second theoretical approach would say that humans have evolved in a manner that selects naturally for prosocial, reciprocally altruistic behavior. The fact that reciprocal altruism can be observed among many animal species strengthens this theory. For example, if a vervet monkey hears another giving a distress call during a fight, it will go to help, especially if the distress caller has helped it in the past. Bonobos share their food and show sympathetic behavior toward disadvantaged members of their group. Jane Goodall, working in Tanzania, observed chimpanzees adopting orphaned infants and sharing meat after cooperative hunts, and female baboons, as mentioned earlier, tend to groom other baboons from whom they themselves have recently received grooming.

Marc Hauser and his colleagues at Harvard carried out experiments with another nonhuman primate, the cotton-top tamarin monkey, to analyze the factors contributing to food sharing among genetically unrelated individuals. They found that a monkey would preferentially give food to a member of its own species who has already been trained to give food itself, regardless of whether it receives a return. In subsequent studies, it emerged that tamarin monkeys altruistically give food to others, but not indiscriminately: they give less food to other monkeys who are selfish and more to those who themselves sometimes give food. Among these monkeys, sharing begets sharing.

This kind of behavior extends far beyond the primates. Among vampire bats, which forage at night, some individuals return to their communal roost without food on any given night. The successful bats share their food with hungry ones; even more impressive, they tend to feed other bats who previously helped them. Long histories of social interaction between these individual animals foster reciprocal altruism. Shared fates allow the altruistic individual to act as though it shares fears, whether of physical defeat (the vervet monkey) or hunger (the vampire bat). Birds, too, are capable of altruism. For example, many small birds unselfishly betray their presence and location by emitting alarm calls to warn the other small birds of an approaching hawk.

Even insects can exhibit altruistic behaviors risky enough to compose a form of socially motivated suicide that Edward O. Wilson compares to a soldier throwing himself on a live grenade in order to save the rest of his platoon. Some ants, bees, and wasps defend their nests by solitarily charging intruders. Other social bees act en masse to attack the hair of a human intruder or to deposit a noxious solution onto the intruder's skin. African termites attack enemies of other insect species with a "yellow glandular secretion that fatally entangles" both the altruistic attacker and his antagonist. All these behaviors put the individual insect at mortal risk on behalf of its group.

How could reciprocal altruism have evolved? The answers are still under debate, but two theoretical points call for consideration. First, among animal species and humans who live in groups, being in the group apparently provides advantages for survival and the raising of young: shared resources, divisions of labor, mutual defense, and so forth. But natural selection works on individuals. Robert Hinde concludes from this that the reproductive fitness of an individual—the chance that he or she will get to pass on his or her DNA to successive generations—depends on the ability of that individual to live happily with others. In the words of Edward O. Wilson, "Compassion is selective and often ultimately self-serving."

A second consideration, one that goes along with the first, focuses on competition and natural selection among groups. Hinde infers that those groups that have been able in the course of evolution to weed out selfish individuals and increase the numbers of prosocial individuals will, as groups, survive better. That is, societies in which most individuals behave unselfishly will last longer and better than societies in which too few people show reciprocal altruism. The first idea, natural selection of the individual, and the second idea, the role of reciprocity in producing effective groups, far from contradicting each other, support each other. Again, Wilson: "Man defending the honor or welfare of his ethnic group is man defending himself."

Hinde quotes evidence that our minds are excellent at detecting cheaters who are breaking a social contract. Moreover, we come equipped with a sense of moral outrage. Thus, with both our highly developed cognitive capacities and our strong emotional dispositions, we are ready to impose sanctions on those who deviate from reciprocal altruism. We even derive pleasure from punishing those who cheat, according to studies painstakingly designed by Tania

Singer and her colleagues at University College London. But the question of how reciprocal altruism has evolved naturally leads to my main question: What are the brain mechanisms by which it works?

Cooperating machines

Powerful support for the existence of brain mechanisms behind reciprocal altruism comes from the realm of machinery. The evolution of cooperation is actually fairly easy to work out, because it follows mechanical steps. Even computers can be programmed to behave in this way. If computers can display cooperative responses, then we know that straightforward physical mechanisms must be sufficient to explain the result. Robert Axelrod and William Hamilton, at the University of Michigan, were able to program computers to display mutual cooperation. Furthermore, by running a "tournament" among computers, they showed that such behavior composes an evolutionarily stable strategy.

Axelrod and Hamilton used the game called "Prisoner's Dilemma" as a vehicle for exploring the dynamics of cooperation. This game, in its simplest form, gives each of two players, A and B, the chance to cooperate with the other or to defect on the other. The story of the game is that A and B committed a crime jointly. They are being interrogated by the police in separate rooms. By not admitting anything to the police—by both protesting their innocence—they can cooperate with each other and go free. Then again, either or both could attempt to get a "squealer's" reward by ratting out the other prisoner, naming him or her to the police. The payoffs for the four combinations of actions in any given round of the game are shown in Figure 1. Note that the selfish choice, defection, has the highest payoff for the individual, but mutual defection has a lower payoff than mutual cooperation. On one hand, it pays each player to defect, regardless of what the other has done. On the other hand, if both players defect, they each get a lower reward than if they had cooperated. That is the dilemma.

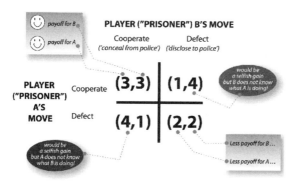

Figure 1. "Prisoner's Dilemma." The game envisions two prisoners being questioned by police. This schematic shows how outcomes can vary: if the suspects cooperate with each other and keep quiet, both have a chance to go free (3,3). If both tell what they know, they can be punished (2,2) but not as severely as one who stayed silent while the other talked (1,4 and 4,1). Hamilton and Axelrod programmed computers to play the game and showed that cooperation can be produced by mechanical processes.

Axelrod and Hamilton emphasize that if the players were never going to play one another again, defection might be the best strategy. However, as in most relations among people, the players will meet again and again in the iterated Prisoner's Dilemma game.

Axelrod has devised computer tournaments that test a wide variety of strategies submitted by mathematicians, economists, sociologists, and political scientists. The simplest strategy, called "Tit for Tat," follows a two-step rule: (1) Cooperate on the first move, and then (2) during subsequent moves, do whatever the other player did on the preceding move. Axelrod calls this strategy "cooperation based on reciprocity." In these tournaments, the strategy of mutual cooperation won the highest scores—that is, mutual cooperation started spontaneously, thrived, and resisted opposition.

Axelrod and Hamilton designed their study to begin with a worst-case scenario, with a population that always acted selfishly. Even in this context, they found at least two mechanisms by which reciprocal cooperation can get started and then thrive. First, if a small number of individuals who are genetically related to each other begin to cooperate, they will continue to do so. They do this because their close relation essentially recalculates the payoff matrix for the game. An individual player has a part-interest in the other player's gain because he is related to the other player. The second mechanism is called clustering. If a small number of individuals in the population have a high proportion of their interactions with each other, calculations indicate they can cooperate to their mutual benefit. Clustering could result from kinship but does not have to do so. Since all these mechanisms work without reference to ethics or religion—they are played out in mathematical games set up for computers—one can be confident they rely on physical mechanisms of the sort that neurobiologists dream of finding.

During computer simulation games that posit complex social settings, the machines adopted strategies that Axelrod called "don't rock the boat." After three mutual cooperations the computers would accept an apology—that is, they would signal their acceptance—in the unhappy case that there had been a noncooperative act. Axelrod's computer simulations show that reciprocity of the behavior of two automata toward each other can be sustained if there is a long enough history of interactions between them.

Here, then, is what I make of this. We are more likely to treat another person properly if we expect to see that person again and again. We will obey this rule if we sense that we inhabit the same "space," the same social domain as the other person. The two of us feel subject to shared fates; we don't harm people with whom we share fates.

Exercises in mathematical game theory such as Tit for Tat and Prisoner's Dilemma have been around a long time, since Merrill Flood and Melvin Dresher (at the Rand Corporation) and Princeton University professor Albert Tucker introduced the latter in the 1950s. Only in the last few years, however, have we had the tools to begin investigating how mutual cooperation depends upon mechanisms in the brain. In 2005, Brooks King-Casas and Read Montague, in the Human Neuroimaging Laboratory at Houston's Baylor College of Medicine, used functional magnetic resonance imaging (fMRI) to observe the human brain while two subjects participated in the repeated economic exchanges of a so-called trust game. Benevolent behavior in this game was associated with higher activity only in a small part of the zone below the brain's cortex that controls movement, the caudate nucleus. Surprising changes in the partner's behavior elicited higher activity in two regions of the frontal cortex as well as in the thalamus and the midbrain colliculi.

It is tempting to conclude that the caudate nucleus is both necessary and sufficient for mutually cooperative behavior. We do not yet know, however, whether this is the case in real-life situations. To find that out will require more complex economic games that can take into account a larger number of choices, greater ambiguity, and longer chains of behavior over time, as well as a greater number of players. The mathematical games explored by Michael Doebeli of the University of British Columbia and his collaborators, for example, have cooperative investments that yield significant benefits to the cooperator as well as to others; the evolutionary dynamics of the game thus become richer and more diverse.

These studies provide some of the first observations pointing to brain regions involved in regulating cooperation and other prosocial behaviors. I believe that a further search for ethics in the brain, so to speak, can lead us to specific and precise mechanisms that come into play when people decide to act in a cooperative, friendly, prosocial manner. It is my hope that the theory proposed in this book will advance this search by suggesting concrete, testable principles potentially responsible for ethical behavior. Presenting my theory will take us on a tour of the neuroscience behind diverse human behaviors and emotions, from fear and anxiety to love, friendship, aggression, and violence.

3

BEING AFRAID

In describing the computer models of cooperation, I mentioned a neurobiologist's dream, that of
finding concrete, physical mechanisms underlying
behavior. Many neurobiologists are pursuing that
dream with the tools of neuroscience, especially
brain imaging, to produce intriguing results such
as the Baylor College finding that linked specific
areas of the brain with benevolent acts.

But the brain has many, many structures and
systems and, to complicate matters by orders of
magnitude, for numerous tasks that we think of
as basic, the brain combines these structures into
elaborate circuits to get a job done. So we must
expect that if ethical behavior is "hardwired," the
brain has some kind of a systems approach, if only
because of the multitude and diversity of human
situations that have an ethical dimension. On the

other hand, if the Golden Rule is so universal, the mechanism behind it must also be rather simple and dependable, and capable of
operating without our having to sit down like The Thinker and cog-
itate on it—the way, for example, sleep is.

These requirements dramatically narrow the candidates for systems capable of generating ethical behavior. I'd like to propose two
such major candidates, depending on whether the ethical circumstances we face have to do with being kind and helpful or with the
possibility of doing harm. Both are far-reaching, both meet the always "on" criterion, and both are very vigorous in the brain in all
our waking moments. One is the system that guides our loving and
friendly behaviors, accounting for love and sociability in all their
forms. The other, the one that (usually) keeps us from harming
others, is concerned with our sense of fear, anxiety, and danger.

Neurobiology is revealing the fear circuits to be a subtle and
multifaceted system, constantly engaged to guide us safely through
our physical and mental environment. The signaling circuitry of
fear works very fast—it needs to do so if it's going to keep us alive.
It registers little things and big things, anything that could mean
danger and the need to protect ourselves in either a minor or a life-
preserving way. And it is tuned in on other major systems in the
brain both to make sure that no danger goes unnoticed and to signal that action is needed in risky situations. To top it off, it is a vast-
ly interesting system, almost a universe in itself. To make fear the
invaluable tool it is, Nature has involved a brilliant collection of
genes, hormones, and structures and calibrated them in such a way
that infinitesimal differences in their operations lead to startling
differences in behavior. And so I find it irresistible to begin with the
fear system.

Think of all the information we have to process in a split second when we're sizing up a possible threat: the shifting shadow
of a stranger in the darkness behind you, a sudden pounding of
footsteps closing in, a spasm of panic as your heartbeat quickens,

a rapid fantasy of punching the stranger in self-defense, and sudden relief, mixed with a tinge of guilt, to recognize your next-door neighbor catching up after having gotten off the bus at the same stop as you did. A massive amount of brain power goes into assessing whatever registers to us as a risky or dangerous state of affairs; thus it is reasonable to think that the brain, which consumes a disproportionately huge share of the body's total energy, would make the best possible use of that energy by applying some of its more elaborate circuitry to more than one job.

Fear can spur the Golden Rule circuitry into action, causing people to opt for ethical choices in a variety of situations. Conversely, when we fail to make a successful response to an ethical challenge, it's possible that something went wrong in one of the elements of the fear system.

The amygdala: headquarters of fear

If you've ever sat in your living room absorbed in a book and suddenly looked up, sniffed the air, and then thought "Fire!" you know the stimuli that lead to fear don't necessarily announce themselves as such from the first microsecond they reach our sensory receptors. Instead, they enter the nervous system value-free, without any emotions attached: the senses of smell and taste through our nose and tongue, the sense of touch through the skin and up the spinal cord, vision and hearing and our sense of balance from our eyes and ears.

We *know* we are afraid only when the fear-producing stimuli reach the brain structure called the amygdala—a remarkable part of our forebrain (which, as its name suggests, consists of all the structures at the front of the brain, on top of the brain stem and the cerebellum). The amygdala plays a major role in controlling a range of negative emotions, from fear and anxiety to anger, stress, and depression. It is a very old structure, which points to its great value as a site that was worth preserving in the course of evolution, and one of its functions is to confer a frightening quality on any input that reaches it. Once the sensory signals get to the amygdala, they acquire the emotional component of fear. (Does this stepwise process suggest that we can voluntarily control fear? This may be an intriguing idea, but the control is going to be only partial at best, as is true for the control of any other human emotion.)

The sense of smell takes the most direct route to the brain, reaching the amygdala almost immediately. This makes evolutionary sense in terms of survival, as it is the sense of smell that provides an advance warning about distant danger. In fact, many of the stimuli that animals associate with fear are smells and pheromonal signals (scents animals release to affect the behavior of other members of their species); both evoke a strong response from the amygdala in common laboratory animals and in higher animals such as primates.

The other four senses send their signals to the amygdala by a roundabout route: up the brain stem (which is actually several structures at the base of the brain where it connects with the spinal cord) toward the front of the brain and into a region called the thalamus. This term comes from the Greek word meaning "antechamber," and the thalamus indeed acts as an antechamber, a relay station for preprocessing sensory signals. In the thalamus, a remarkable thing happens. Sensory signals now bearing the red flag of fear go off in two separate directions, some directly to the amygdala and others to the magnificently wrinkled outer layers of our brain, the cerebral cortex. These two streams of information behave in different ways: according to Joseph LeDoux, at New York University, inputs from the cortex are huge at first but tend to fade over time, whereas inputs directly from the thalamus are smaller but remain stable over time. The cerebral cortex apparently distinguishes between various types of fearful information in a way the thalamus itself does not; in any case, both sites then send their processed information down to the amygdala.

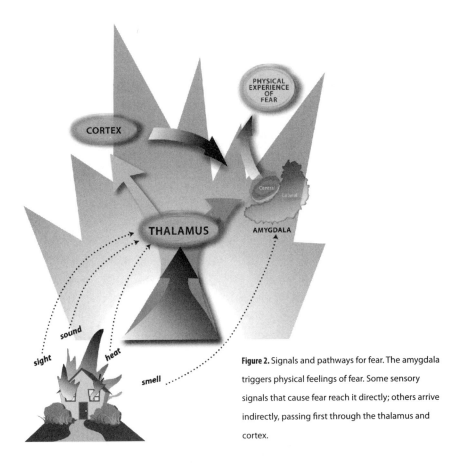

Figure 2. Signals and pathways for fear. The amygdala triggers physical feelings of fear. Some sensory signals that cause fear reach it directly; others arrive indirectly, passing first through the thalamus and cortex.

While smell reaches the amygdala directly, another kind of sensory signal to get there is sight. Belinda Liddell and Leanne Williams, at the University of Sydney in Australia, have reported that patients who have suffered damage to the part of the brain that houses the visual system cannot "see" frightening stimuli; that is, they are not conscious of noticing the appearance of such stimuli, as you or I would be, and cannot describe these sights. Nevertheless, they react to them much as we would: with fear. One way to interpret these results is with the possibility that frightening visual stimuli activate the amygdala directly from the brain stem, without involving the visual cortex. To test whether this might be the case, Liddell and Williams showed their participants a series of frightening pictures, but presented each one so briefly that they registered on the participants only subliminally—that is, beneath the threshold of consciousness. The scientists found that in just milliseconds, these frightening stimuli activated visually responsive areas in the brain stem and, most importantly, in the amygdala, without stirring up a response from the visual cortex.

Just as sensory signals give rise to fear, so fear in turn (even unconscious fear) feeds back into our sensory experience. A research team led by Beatrice de Gelder and Ray J. Dolan, at the Institute of Neurology in London, studied a patient who experienced nonconscious "vision" in one visual field—he responded to visual stimuli in that particular field even though he could not consciously see them. When the researchers presented a frightening image to this patient's nonconscious visual field, he reacted with fear as anyone would; apparently the frightening stimulus worked on him through its effects in the amygdala and the cortex.

More support for the amygdala's central role in fear comes from studies in monkeys. At the University of Wisconsin, Richard Davidson and Ned Kalin have used rhesus monkeys, which are afraid both of snakes and of human intruders, to study the mechanisms of fear. The scientists injected a small dose of a toxic chemical into the portion of the monkeys' amygdala called the central nucleus, destroying this group of nerve cells. As a result, under the same circumstances that had previously made them display fearful behavior, the monkeys now appeared significantly less afraid than before the injection.

While almost all the scientific work linking the amygdala to fear has been done by careful experimentation on laboratory animals, a few studies with human volunteers have helped to fill in the picture. Elizabeth Phelps and Joseph LeDoux have used functional magnetic resonance imaging (fMRI) techniques to find that humans, too, are subject to fear conditioning—a learned fear response we will discuss in greater detail in the next chapter—and that, as with animals, this process takes place in the amygdala. These studies showed that even when fearful stimuli acquire their emotional property through verbal instruction, they still work through the amygdala.

Another study of the human amygdala reveals this brain structure's central role in people's negative emotions in general, including fear. Stacey Schaefer and Sharon Thompson-Schill, at the University of Pennsylvania, told volunteers to hold onto negative feelings—to keep in mind something that upset or distressed them—while they underwent brain imaging. The images showed that the volunteers' negative emotions corresponded closely, in time and in their intensity, with levels of activity in the amygdala. Hence the Pennsylvania research team inferred that the negative emotions were due at least in part to this activity. All these results indicate that what we have learned about how the amygdala handles fear in the brain and behavior of animals applies equally well to humans.

Can the amygdala be thought of as a "headquarters" of fear? While this image is basically correct, it needs to be modified in light of the current way of thinking about the function of different brain regions. In the older days of brain research, experts thought in terms of "centers" in the brain. They supposed that each center would control a specific function: recognizing emotions on other people's faces, analyzing the sounds of speech, and so on. Now, instead of centers, we talk about circuits through which information passes. A unique circuit, with various points at which signals from different parts of the brain come together or diverge, would explain quite well how a specific behavior or feeling comes about. Thus, the amygdala can be viewed not as a fear "center" but as a crucial node in fear circuitry. Moreover, we now know that distinct groups of nerve cells in the amygdala represent crucial nodal points or junctions in the circuits that contribute to the feeling of fear, the memory of fear, and even the suppression of fear. These circuits then carry signals to many outputs in other parts of the forebrain and midbrain.

In a couple of ways, though, the amygdala (named for its "almond-like" appearance by the late, great neuroanatomist Alf Brodal) is proving a tough nut for scientists to crack. The first set of complications comes from its highly complicated structure: the amygdala comprises more than ten distinct cell groups, and each cell group transmits and receives its own distinct signals.

Further complications arise from the blurriness of the physical borders between the assorted cell groups; it is hard to manipulate one group in an experiment without affecting another inadvertently. Some help may come from new techniques such as laser capture microscopy, which allows researchers to pluck out an individual neuron from a cell group under microscopic control and to measure the amounts of the various proteins produced by its genes. We can even record the electrical activity of individual *parts* of neurons within brain tissue slices through a cell group of the amygdala. Thus, we are advancing to greater and greater mechanistic detail, from brain regions to cells to the genes active within those cells that contribute to the mechanisms of fear.

The amygdala, however, does not account for all aspects of fear. All our apprehensions fall into two categories: primitive emotional reactions and more sophisticated cognitive, thoughtful reactions. On the basis of what I've read in the current scientific literature, I surmise that the *emotions* of fear come straight from the amygdala itself; similarly, innate, unlearned fear originates in the central nucleus of the amygdala; however, our more complex fearful *thoughts* arise from the information-processing power of our cerebral cortex, which I will discuss in greater detail in the next chapter.

Producing fear one step at a time

What events take place in the neurons, or nerve cells, of the amygdala to create the feeling we call "fear?" When you are frightened by approaching footsteps or an alarming smell, this means that your brain's signaling pathways have hurtled through at least three main steps: the arrival of specific chemical signals at the amygdala's neurons, the signaling within these neurons after the arrival, and the eventual effect of that signal (which may be either an electrical discharge from a neuron or a change in gene activity).

Among the major types of signaling molecules operating in the brain are chemical messengers called neurotransmitters and protein fragments called neuropeptides. After the arrival of particular neurotransmitters and neuropeptides at a neuron in the amygdala and the chemical signaling within that neuron, it is the third step that determines the consequences of the signaling, which can be either fast or slow.

The fast effects take place in the membrane that encloses an amygdala neuron, and they result in a rapid change in the neuron's electrical activity. This change follows a "decorative" molecular event called phosphorylation, in which portions of neurons are "adorned" with appendages containing phosphorus and oxygen. In the signal circuitry of fear, phosphorylation affects a protein that is part of an ion channel in the nerve cell membrane, and that channel is likely to be one that permits the flow of calcium into the cell or the flow of potassium out of the cell.

The slow effects of the fear signaling, taking place in the cell nucleus, result from the phosphorylation of DNA-related proteins. These effects account for slow, gene-related adaptations that make for long-lasting states of fear. When such prolonged states develop over a matter of hours, certain genes become activated and turn out a variety of proteins related to stress, the immune system, and other body functions. The mechanisms behind these two types of effects—the rapid-electrical or slow-genomic—might one day provide possible targets for future drugs that would reduce fear in a therapeutic manner.

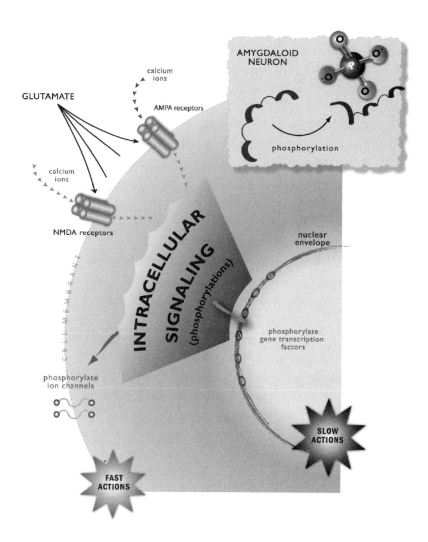

Figure 3a. Glutamate and fear. Glutamate binds to NMDA and AMPA receptors on amygdala neurons to promote fear. Certain aspects of fear depend on which receptor glutamate binds to; blocking this binding might be a way to reduce inappropriate fear.

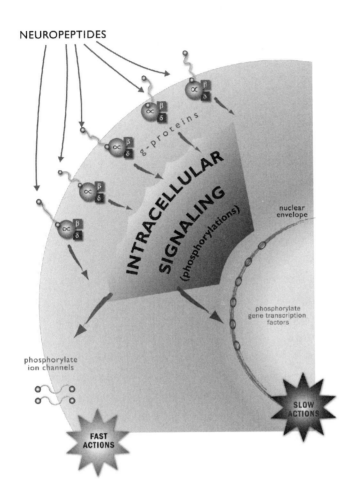

NEUROPEPTIDES

g-proteins

INTRACELLULAR
SIGNALING
(phosphorylations)

nuclear
envelope

phosphorylate
gene transcription
factors

phosphorylate
ion channels

FAST
ACTIONS

SLOW
ACTIONS

Figure 3b. Neuropeptides and fear. Neuropeptides influence cells in various ways, including stimulating signaling. A neuropeptide called BDNF may participate in fear "conditioning," or learning to fear.

Glutamate

One rapidly acting neurotransmitter important for the production of fear in the amygdala is glutamate. It has a long evolutionary history—this transmitter was one of the first to come into existence in animal brains—and it binds to many different kinds of proteins called receptors. Such binding is at the core of all chemical reactions in the body: a chemical messenger and its receptor are said to fit each other like a key and a lock, and once the key enters the lock, it unleashes the chemical reaction in question. However, one "key," such as the chemical messenger glutamate, can fit into different receptor "locks," and the resultant chemical reaction will be determined by this binding.

One type of receptors to which glutamate binds on the neurons of the amygdala, called the AMPA receptors, is distributed in a particular pattern. Studies by Roberto Malinow and his colleagues at the Cold Spring Harbor Labs tell us that the location of the AMPA receptors on these neurons is crucial for fear conditioning, determining how effectively such conditioning takes place.

Two research teams, one led by Michael Davis at Emory University and the other by James McGaugh at the University of California at Irvine, have had the insight to block glutamate receptors as a way of reducing inappropriate fear. McGaugh and his colleagues used a chemical that blocks a different kind of glutamate receptor from the AMPA receptors I mentioned above. After this chemical was infused into the lateral amygdala of rats, the animals' ability to hold onto a fearful memory was significantly weaker than that of a control group.

Inspired by these findings, Davis has searched for the kind of drug that would assist psychotherapists in their attempts to reduce patients' phobias and stress disorders. He has come up with a compound called D-cycloserine, a chemical that helps to extinguish frightening memories. D-cycloserine binds to a special type of glutamate receptor, and Davis theorizes that when injected into the amygdala, it displaces the normal, more effective substances that belong there and that are responsible for fearful memories. Davis put his theory to a test in a clinical trial with patients who have acrophobia, a fear of heights: he and Kerry Ressler used psychotherapy in combination either with D-cycloserine or with a placebo drug (a dummy pill). The reduction of fear, as manifested by the patients' behavior as well as by physiological measures, was significantly greater in patients receiving D-cycloserine than in those receiving the placebo. This research provides yet another demonstration that in producing fear, the role of glutamate receptors in the amygdala is paramount.

Calcium

The effects of glutamate on neurons of the amygdala depend on a chain of chemical reactions that involves calcium. By binding to its receptors, glutamate allows the ion channels of the neuron's membrane to open, and the open channels permit calcium to enter the cell. Calcium, in turn, unleashes a remarkable cascade of signals in the cytoplasm of the cell, outside its nucleus. This cascade makes use of special kinds of enzymes, called kinases—proteins facilitating such biochemical reactions as phosphorylation, the "decorative" molecular event mentioned above.

The entry of calcium into the neurons of the amygdala and the resultant cascade of signals are of central importance for the memory of fear, as has become clear from several lines of evidence. For one, we know that a certain kind of calcium channel is present in a particular type of neuron in the lateral amygdala. This is the L-type calcium channel, the same sort that plays a major role in the functioning of the heart, and it's found in the membranes of large, pyramid-shaped amygdala neurons that are essential for fear learning.

Another line of evidence has to do with kinase enzymes that are part of the described above calcium-dependent cascade. Studies indicate that neurons of the amygdala contain a key kinase enzyme of this sort, and research performed by New York University's LeDoux reveals that during fear conditioning this enzyme is being phosphorylated, as it should be in the cascade. In fact, if scientists inject a chemical that blocks the action of this kinase into the amygdala, the experimental animal becomes unable to learn a new fear memory, even with special training.

The third line of evidence represents a middle ground between the other two. Transgenic mice that had been genetically manipulated to eliminate a gene encoding one of the calcium-dependent kinases (so that they had some kinase, but not as much as normal mice) showed a significantly lower ability to remember frightening experiences. All these findings point to both calcium signaling and a chain of calcium-dependent enzymes as being crucial for fear.

BDNF

Another chemical messenger important for fear is a neuropeptide called the brain-derived neurotrophic factor, abbreviated as BDNF. The main role of BDNF is to make nerve cells grow, especially those connected with the autonomic nervous system (which handles routine matters like sustaining a steady heartbeat in your chest as you read this). However, BDNF is also known to take part in the circuitry of stress and fear. Experimental work that disrupts the signaling of BDNF in the amygdala of laboratory animals seriously interferes with fear conditioning. Genetic mutations that block the genes' synthesis of BDNF likewise impair some forms of fear conditioning.

The influence runs in both directions; fear alters the activity of BDNF genes themselves. In one study, Davis and his colleagues at Emory University kept a close watch on the activity of the gene encoding for BDNF in the amygdala; they found that two hours after fear conditioning, this activity had increased in the amygdala. The scientists also discovered a clever regulating mechanism involved in this activity: the BDNF gene can be turned on by four different DNA segments, called promoters, and only one of these promoters participated in the fear circuit. An assortment of promoters for an individual gene allows considerable flexibility and variety of genetic regulation, and in the case of BDNF, too, they probably allow this neuropeptide to perform different functions in the brain.

Body language of fear

When we respond to fearful stimuli, the brain rapidly shifts signal-processing energy away from the steady monitoring of "normal" stimuli to the vital challenge of identifying the danger, if any, that faces us. This shift of signal-processing resources is why background sights and sounds seem to fade away at moments of crisis; in an action film, the hero may make his death-defying leap from the airplane in silence and in slow motion to narrow down our senses to a pinpoint focus on that one instant.

How do the fear-related signals produce the overall state that engulfs both our mind and body? Where do these signals go from the amygdala?

To explain both the emotional and the physical components of fear, let us concentrate on the signals the amygdala sends to parts of the brain that control our visceral responses—literally, our gut reactions. Consider the mingled panic and rage of a conscientious worker who is suddenly dismissed from his job; or a violent fit of a betrayed husband who is seized with the mixed emotions of rage and fear of being abandoned. A brain circuit carrying signals from the central nucleus of the amygdala to the septum—a structure in the midline of the forebrain—would explain just this kind of emotional reaction, which is known as "septal rage." Unlike ordinary anger, septal rage is sudden and vicious. Such a circuit has been identified in animals under experimental conditions. However, the signals the septum transmits are exactly the opposite of what one might expect: these signals actually reduce the level of activity at several brain sites involved in the neural circuitry of aggression and fear. In other words, this is a brain circuit whose job is to prevent behavioral meltdowns. Laboratory animals whose septum has been damaged demonstrate rage. As for the urge to strike back at a source of fear, that comes from the hypothalamus, a cherry-sized brain region, located, as its name suggests, below the thalamus (we will encounter this brain region on numerous occasions because in addition to regulating such functions as body temperature and the release of hormones, it is involved in the control of fear, stress, and loving responses). The urge to strike comes from the part of the hypothalamus called the ventromedial nucleus, which also receives signals from the amygdala. Damage to this cell group leads to vicious aggression, but when this part of the hypothalamus functions properly, its level of activity is lowered by the signals from the amygdala.

One more signaling pathway contributes to the feeling of fear: some of the longest nerve fibers from the amygdala reach almost the entire length of the brain, to an area called the locus coeruleus (literally, "dark blue place"). When a frightening event brings us from a soporific or calm state to a state of high alertness, the neurons of the locus coeruleus are likely the cause. This is because locus coeruleus neurons serve to drench the brain in norepinephrine, a neurotransmitter that arouses us quickly.

And what about the sweating, shallow breathing, and rapid heartbeat of fear? Some neurons in the amygdala send axons (long projections) to an as yet poorly understood site in the very middle of the brain that consists of nerve bodies and therefore is known as central gray. If the results from animal research in this area hold true in humans, these neuroanatomical connections would account for our bodily reactions during our feelings of fear.

Thus, signals from the amygdala produce the emotional body language of fear, and this language appears to reflect each person's individual temperament or character. The laboratory of Richard Davidson, at the University of Wisconsin, is using state-of-the-art brain-scanning techniques to link mental activity to the changes in heartbeat that come with anxiety and fear. Davidson and his colleagues exposed young adult volunteers to stimuli that either predicted a

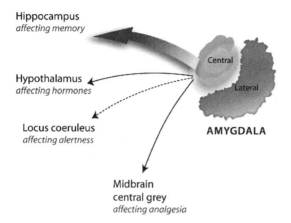

Hippocampus
affecting memory

Central

Lateral

Hypothalamus
affecting hormones

Locus coeruleus
affecting alertness

AMYGDALA

Midbrain
central grey
affecting analgesia

Figure 4. Sending the message. After fearful stimuli activate amygdala neurons, the message "be afraid" is distributed to other brain regions in order to coordinate our responses.

painful shock or gave an "all-safe" signal. They found not only that the amygdala lit up when the volunteers were experiencing fear or shock, but also that the level of activity in the amygdala was correlated with the level of cardiac activity in response to fear. What Davidson and his colleagues have discovered may prove to be the bodily form of a personality feature, a strongly anxious reaction to threat that varies in different people depending on their personalities.

In studies with two-year-old children, Davidson and Kristin Buss, then at the University of Missouri-Columbia, used a milder but still threatening context to elicit fear. They concentrated on children who had extremely fearful reactions—that is, those who easily show distress to fearful stimuli. The scientists found that, when in distress, the children's tendency to freeze in place in mid-task was tightly correlated with their bodily reactions such as speeded-up heartbeat and elevated levels of stress hormones. Since those bodily reactions were still evident a full week after the upsetting experiment was over, I assume that here, too, the children's individual characteristics came into play: the physical signs that Buss and Davidson observed could be real, physiological manifestations of the children's personality, in terms of their capacity for showing distress.

Primed for fear

All the frightening emotions and memories will not stick with us as they should—that is, in a way contributing to our survival—if the entire central nervous system has not been adequately prepared for the task, or, to use a neuroscientific term, "aroused." By coordinating activity among several different areas, the neuronal circuits of arousal enable us to become alert to stimuli that are surprising, out of the ordinary, perhaps even dangerous: By making our heart thump strongly and our breath come quickly and shallowly, the brain readies the body to send floods of extra energy to our muscles in case we will need to act fast—by leaping up and shouting "Fire!" or at least rushing to the kitchen to see if something is burning.

McGaugh and his colleagues at the University of California at Irvine have observed that the proper working of fear-related mechanisms in the amygdala depends on nerve signals that release two neurotransmitters involved in brain arousal, norepinephrine and dopamine. For example, in 2005, Ryan LaLumiere and McGaugh reported on a study in which they trained rats to avoid returning to a place where they had undergone an electric shock to the feet. The researchers found that when they infused the animals' lateral amygdala with a chemical compound that prevents dopamine from binding to its normal receptor sites, the infusion prevented the rats from acquiring fearful responses as well as they otherwise could. Conversely, infusing dopamine itself or, for that matter, norepinephrine, into the lateral amygdala of the same animals helped the animals to retain their fearful responses. Additional shocks between training and testing—enough to arouse their central nervous system—also increased their subsequent fear responses.

Thus, animals with lower levels of generalized brain arousal are less likely to show high levels of learned fear responses. This probably corresponds to our human experience of the proverbial un-flappable "cool guy" who does not seem to register fear during situations that truly deserve a full-blown fear response. On the other hand, the opposite is also true, namely, a person in the throngs of a full-blown fear response is probably not going to look too "cool." That is due, in part, to the fact that an adequate fear response, in turn, appears to raise brain arousal: according to the University of Sydney researchers Liddell and Williams, fear-related signals could go from the amygdala to the arousal systems of the brain.

These insights into an effective fear response, as well as all the other mechanisms described in this chapter, explain how fear is produced and what kinds of effects it generates in the body. The next step in linking fear to the observance of the Golden Rule, as proposed by my theory, requires examining how the fear system fits in with other important systems and functions of the brain and body, including those that support memory, learning, and response to stress.

4

MEMORY OF FEAR

Once bitten, twice shy. This popular saying aptly sums up the impact of memory—such as the memory of fear—on our behavior, the crucial value of past knowledge for future behavior choices. A repertoire of memories of what fear, danger, injury, threat, unhappiness, and betrayal feel like helps us decide how to act in various situations, including those that pose an ethical dilemma. One of the formulations of the Golden Rule impels us to refrain from doing unto others what we might not wish done to us, which, in turn, implies being able to remember whether or not we'd want something to happen to ourselves. Therefore, to reveal how ethical behavior comes about, we need not only to understand the mechanisms of fear that I described in the preceding chapter, but also to clarify how we learn to fear something and hold on to this memory so that it can serve us in case of need.

Learning to be afraid

We have two kinds of fear: fear that we learn explicitly (such as realizing that when a parent suddenly calls you by your full name, you are probably in trouble), and fear that is innate, or unlearned. Some people seem to be innately afraid of snakes; others are afraid of heights, and still others of flying, regardless of whether they have ever experienced a snakebite, fallen from a great height, been in an airplane accident, or, for that matter, engaged in studies into the brain mechanisms of fear. New York University scientist Joseph LeDoux, whose studies I often cite in connection with fear research, himself is afraid of snakes; the famous football announcer John Madden refuses to travel by plane.

What is the neurological basis of learned fears? As mentioned earlier, many fears are "conditioned," which means that an animal or human learns to respond with fear or alarm to a stimulus that is not frightening in itself. As shown by LeDoux and his colleagues, this kind of learning takes place when a stimulus that is originally neutral—say, red hair—becomes associated in the individual's brain with unpleasant events—for instance, back in third grade, the carrot-topped kid was a bully who made you his favorite target. In such a case, conditioned fear may be the reason that to this day you still avoid redheads. In small laboratory animals, such as mice or rats, if the sound of a gong is consistently followed by a brief electric shock to the foot, the animals soon come to react with fright, freezing in place, when they hear the gong, even when the shock fails to arrive. (Such "freezing" stems from an innate fear response that in nature no doubt saves their hide from time to time; a stiff immobility would, for example, make small animals harder to detect against a visual background and would also prevent them from making sounds that could be heard by a predator.)

It is hardly surprising that the same brain structures we saw involved in producing fear—and first and foremost, its "headquarters," the amygdala—are also critical for acquiring a fear response.

For people as well as for animals, learning to fear something (or someone) requires the help of the amygdala. Antonio Damasio and his colleagues, at the University of Iowa School of Medicine, have reported that a patient who suffered damage to the amygdala in both the right and the left hemispheres had trouble recognizing fearful expressions on people's faces, specifically because she failed to look at the eyes of the fearful person. Most of us, when we are afraid, instinctively widen our eyes, and the amygdala apparently allows the observer to pick up on this widening. In a study that Paul Whalen and colleagues at the University of Wisconsin conducted on people's reactions to the facial expressions of others, fearful eye whites activated the amygdala of the observers much more strongly than "happy" or neutral eye whites.

In this connection, one cell group that has received a lot of attention over the years is a certain part of the amygdala. This cell group is called the lateral nucleus, sometimes referred to as the "lateral amygdala." Michael Fanselow, of the University of California at Los Angeles, wanted to find out whether the lateral nucleus only helps to form the memory or plays a permanent role in the subsequent consolidation of a fear memory, an important step in preserving the memory over the long term. He and his colleagues worked with rats, pairing a certain tone with a slight but painful electric shock to the feet. When the scientists tested the animals again after an interval of sixteen months, a long time in the life of a rat, they found that even after so much elapsed time, destroying the lateral nucleus significantly reduced the rats' conditioned fear responses. Fanselow concluded that neurons in the lateral nucleus of the amygdala are necessary for the permanent storage of fear memories. These neurons, according to LeDoux, "predict and perceive the aversive qualities" of unpleasant or hurtful stimuli, of the sort that learned fear responses are meant to avoid. In the words of James McGaugh, of the University of California at Irvine, the amygdala "renders emotionally significant experiences memorable."

The amygdala sends its emotionally loaded signals to many other parts of the brain associated with memory—in particular, the hippocampus (see Figure 4, on p. 37). When Christa McIntyre and McGaugh stimulated neurons in the lateral amygdala of the rat, the animal's brain showed increases in the level of proteins associated with the process of memory formation. Conversely, infusing the amygdala with a chemical that interferes with the forming of frightening memories caused a drop in the levels of that same protein in the hippocampus.

Damage to the amygdala can reduce fear conditioning. Joseph Paton and Daniel Salzman, in the Department of Psychiatry at Columbia Medical School, have explored the role of the amygdala in this process. The researchers used microelectrodes to record electrical activity from individual neurons in the amygdala of rhesus monkeys. The experiment had two stages. First Paton and Salzman trained the monkeys to know which of several stimuli would be followed by a treat (for example, a small drink of juice) and which ones warned of an unpleasant surprise (such as a raucous noise indicating no rewards for the next 30 minutes). Then the scientists abruptly changed some of the positive stimuli to negative, and vice versa. After a short period of confusion, the monkeys once again responded appropriately to both positive and negative stimuli. They were able to do this because the electrical activity of the neurons in their amygdala changed quickly to reflect the new values. However, the Columbia researchers found that animals that undergo injury to the amygdala show reduced fear conditioning—they have difficulty learning to be afraid of stimuli that they actually should fear.

When it comes to the more conscious fear responses, the amygdala no longer takes center stage and a prominent role belongs to a network of brain regions centered in the front of the cerebral cortex. They are the regions responsible for the fears that we each learn through our life experiences. According to a brain-imaging study by Luiz Pessoa and Srikanth Padmala at Brown University, these brain regions are responsible for our conscious recognition of a fearful stimulus. They do not handle sensory signals; instead, they regulate our emotional responses and prepare us to deal with emotionally challenging stimuli.

But learning and "conditioning" make up only one side of a coin in the currency of fear. The other side of the coin is learning to suppress fear—something the brain must have a way to do, if our sense of fair play is ever to dominate in ethically challenging situations.

How fear fades

Once afraid, we cannot stay afraid forever. Fortunately for my theory of fair play, the fear system is equipped to help us let go of fear when that is appropriate. (It's fortunate for our bodies, too: a chronically clenched gut and pounding heart would be a very uncomfortable state in which to live.) If someone who matters to us is in danger, we need to be able to put aside our instinct for self-preservation long enough to help—and this brain circuit, which presumably evolved eons ago for parents to protect their children, would be available in all our brains nowadays for use by beneficent strangers who rush to the rescue in drownings, building fires, and all kinds of calamities.

A special region at the front of our cerebral cortex, the prefrontal cortex, helps us to manage fear—to suppress or "unlearn" it when necessary. In the early days of neuroscience, experiments with monkeys showed that damage to the prefrontal cortex harmed the animals' ability to "unlearn" their fearful responses. This early finding turned out to be important for our understanding of fear conditioning. Solid scientific evidence since then shows that the brain has two distinct types of circuits: for learning that something is dangerous—in other words, it is to be feared—as well as for learning that something is *not* dangerous after all, and that key sites in the latter circuit are located in the prefrontal cortex.

Studying the rat brain, Maria Morgan and LeDoux found that damage to a particular portion of the prefrontal cortex, the ventral medial prefrontal area, disrupted the animal's ability to "unlearn" a conditioned fear. The scientists inferred that the role of this area under normal circumstances is to tamp down fear in response to a conditioned stimulus when it no longer represents a threat. Even if the animal had received training to inhibit that fear, rats with damage to the prefrontal cortex were stuck with greater levels of fear than those without such damage.

Even temporary damage to this fear-suppressing circuit may make it impossible for the animal to stop being afraid of a conditioned stimulus. Mark Laubach, of Yale University, injected a chemical that mimicked the action of an inhibitory neurotransmitter for a defined period of time into this brain area in laboratory rats; the injection released the animal from the fear suppression and caused premature and inconsistent fearful responses. This research also shows that suppressing fear is rather hard work for the brain: The longer the animal is supposed to inhibit the fear response it so painfully learned, the harder the challenge of suppressing a fearful memory, and the worse the animal performs at it.

The observations gleaned from animal experiments also serve to predict what happens in the human prefrontal cortex. Scott Huettel and Michael Platt, at Duke University, used brain imaging to show that people who behave more cautiously than others have higher levels of activity in their lateral prefrontal cortex. The Duke researchers interpret their results as suggesting that the prefrontal cortex inhibits impulsive, emotionally driven responses.

The prefrontal cortex can probably inhibit fear, at least in part, by countering the activity of the amygdala. Certainly, neurons in the prefrontal cortex can increase or reduce activity in the amygdala itself. Moreover, prefrontal cortex neurons send their long extensions to some of the same places as do amygdala neurons in the forebrain (including the hypothalamus and midbrain), thus provid-

Prefrontal cortex

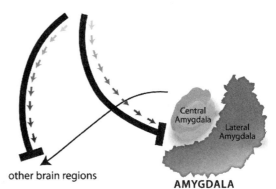

Figure 5. Suppressing fear. Neurons in the prefrontal cortex have the capacity to suppress fear responses.

other brain regions

Central Amygdala

Lateral Amygdala

AMYGDALA

ing additional routes by which prefrontal inhibition could "cut off" amygdala-generated fear responses. An account of how the prefrontal cortex inhibits inappropriate fear responses is a scientific work in progress, and a full picture of this inhibition is yet to emerge.

Apparently the prefrontal cortex controls fear, in part, by establishing a state of expectancy that an emotional event will occur. Felix Bermpohl and colleagues at Harvard Medical School used functional magnetic resonance imaging to measure brain activity when volunteers were "set up" with cues that would lead them to expect either emotionally laden stimuli (pleasant pictures, such as those of a pair of cuddling bunnies, or unpleasant ones, such as a suicidal man with a gun) or neutral stimuli (such as a picture of the same man holding a hairdryer). In the prefrontal cortex, as in the amygdala, expectancy had an amplifying effect: it heightened the neural activation caused by emotionally evocative pictures but not by neutral ones. The greatest increases in activation came when the volunteers were given expectancy cues before seeing the emotionally evocative pictures. The existence of this expectancy stage leads me to speculate that during the suppression of fear, the fear-inhibiting circuits of the prefrontal cortex are primed even before the frightening stimulus is presented.

Cortex control: left versus right

In suppressing a learned fear response, as in controlling our general emotional state, the left and right parts of the prefrontal cortex have not been created equal. In fact, I have been surprised by an important difference between the emotional valence of the two sides of the prefrontal cortex. Most left-right differences in the brain involve sensory, motor, and certain cognitive functions, but numerous recent studies also demonstrate that the two sides of the prefrontal cortex support opposite states of mind. Higher activity in the left prefrontal cortex produces a sense of well-being; higher activity in the right prefrontal cortex is more often associated with misery. It is as if subjects with right-side activation were choosing to look at facial pictures illustrating disgust, whereas those with left-side activation were looking at happy faces. Depressed patients have less prefrontal activity in the left hemisphere than in the right.

This emotional asymmetry has important consequences for the suppression of fear by the prefrontal cortex. Signals sent from the left prefrontal cortex not only reduce the fear but also lighten a person's mood, while signals sent from the right prefrontal cortex, even if they alleviate the fear, can be expected to worsen the mood.

The left-versus-right difference probably also shows up in the effects of the prefrontal cortex on the amygdala, as well as the other brain regions to which the amygdala sends its signals. The signals sent from the left prefrontal cortex most likely activate some of the inhibitory neurons in the amygdala, thus reducing fear and anxiety. The signals sent from the right prefrontal cortex can be expected to enhance neuronal outputs from the amygdala, thus worsening fear. The same left-right asymmetry holds true for nerve cells of the hypothalamus and the midbrain, through which amygdala-produced states of fear have their bodily effects. I would predict that the neurotransmitters used by the neurons of the left prefrontal cortex are different from those used by the neurons of the right, and that the left must be either more effective at inhibiting or less effective at enhancing the fear-generating forces that act on our bodies as a whole.

The leading authority on these prefrontal cortical left/right asymmetries, Richard Davidson, at the University of Wisconsin, conducted an experiment in which volunteers chose words from pairs that differed in their emotional tone. People with a greater amplitude of electrical activity in their left prefrontal cortex were more likely to choose the pleasant words than individuals who did not have left prefrontal dominance. While asymmetries showed up at sites in the central and posterior cortex as well, only the ones in the prefrontal cortex were significant.

Accordingly, it should not be surprising to learn that damage to the right prefrontal cortex—for example, by a stroke or injury—can actually improve mood; it can even be associated with the sense of an emotional "high," whereas damage to the left prefrontal cortex can bring on a chronic case of the blues. In such cases, medicine to step up the activation of the left prefrontal cortex could offer a way to help the patient feel better.

How do we account for such unevenness of brain activity in a feature as important as overall state of mind? I believe that the explanation can be fairly simple. First we have to stipulate that the two sides of the prefrontal cortex send their signals to all the same sites throughout the brain—and that their nerve fibers oppose each other. All it would take in order for the effective neurochemical outputs of the two to be different, or even opposite, would be for different genes to be activated in the neurons of each side.

The activity of genes could also explain the different roles of the left and right prefrontal cortex in extinguishing fear. To examine this possibility in a rigorous way, I suggest a scientific study in which the activity levels of all genes in the left prefrontal cortex are compared with the activity levels of the genes in the right prefrontal cortex. The experiment should be conducted exactly at the point in a behavioral experiment when we would expect the prefrontal cortex to extinguish a learned fear response. The genes activated differently in the two hemispheres will tell us how the two sides of the prefrontal cortex work differently to suppress fear—in one case (the left prefrontal cortex) leading to a placid sense of well-being and in the other case (the right prefrontal cortex) leading to chronic depression and distress.

Traumatic circuitry

What happens when fear fails to fade? Beyond the possible ethical consequences that could be predicted by my theory, which links the activity of the fear system with a potential ethical fallout, this is an important health question. A successful management of fear by the prefrontal cortex is crucial for our mental health because enduring states of panic are not good for us.

On a spectrum of human states of distress, one of the furthest points must be post-traumatic stress disorder, or PTSD. This grievous condition plagues patients who have not recovered from assault, natural disaster, or war—in many ways, their fears refuse to go away. For someone with PTSD, even a trivial perception like the sound of rain striking a window can unleash a torrent of fearful memories, trapping the person into reliving the worst moments of his life over and over.

Brain-imaging studies have reported abnormal degrees of activation in both the amygdala and the prefrontal cortex of people with PTSD, consistent with what we know about the neural circuitry for fear and its suppression. In a study of sixteen Vietnam veterans that Israel Liberzon and his colleagues at the University of Chicago reported in 2006, veterans with PTSD were found to exhibit altered neuronal responses to emotionally salient stimuli, not only in the prefrontal cortex but also in the amygdala. The veterans with PTSD failed to activate the prefrontal cortex in response to very unpleasant pictures, and in that respect they were different from other Vietnam vets who did not have PTSD. A report from the laboratory of John Morrison, of the Mount Sinai School of Medicine, suggests possible cellular mechanisms that could account for these results: the report's findings that repeated stresses can damage the connection-forming structures on the extensions of neurons in the prefrontal cortex could have something to do with that brain region's loss of efficacy in reducing fear.

Changes in the amygdala itself could also play a role. Injecting a biochemical that mimics an inhibitory neurotransmitter into the amygdala blocks the ability of stress to impair memories. This finding suggests that the amygdala facilitates the stress-related memory impairment that can be a blessing in traumatic situations, as such an impairment might help avert PTSD.

Of course, not everyone is equally susceptible to PTSD, and past experiences play a role. UCLA's Fanselow has stated that, in animals, an initial exposure to painful shock appears to increase the creature's fear later on in response even to less stressful experiences. Perhaps among all the people present during an extremely horrible event such as the collapse of the World Trade Center buildings in 2001, those with a history of chronic stress may have been more likely to fall prey to PTSD than those without such a history.

The failure of fear to fade in PTSD reveals a fascinating interplay between two seemingly opposed mechanisms. In a recent scientific book, *Brain Arousal and Information Theory,* I portray experiences that arouse the nervous system, priming it for action, as beneficial on the whole, because of their positive effects through activating neural and behavioral responsiveness. Yet research on PTSD shows a potential role for brain arousal in the *inhibition* of certain responses. In other words, contradictory as this might seem, adequate arousal might be needed for an effective suppression of fears.

As an example, one arousal transmitter in the brain can bring about the inhibition of responses—specifically, for our purposes, fear responses—in at least two different ways. This tiny protein, called orexin, is known for its arousing effect on the nervous system, but several studies have now suggested that orexin is involved in the ability of the prefrontal cortex to suppress inappropriate fear. Specialized techniques for recording electrical activity in the brain reveal that crucial neurons in the prefrontal cortex respond positively to orexin. Moreover, these neurons respond particularly strongly to a combination of orexin and the excitatory neurotransmitter glutamate, thus giving evidence of a seeming paradox: the potential power of a brain-arousing chemical to inhibit certain kinds of behavior. And the evidence comes not only from recording from the cortex activity itself, but also from neurons in the thalamus, which send signals to the prefrontal cortex. Anthony van den Pol and his team at Yale Medical School have reported that these thalamic neurons use glutamate as a transmitter and that they themselves are excited by orexin. In other words, cognitive and emotional arousal could actually help in the prevention of inappropriate fear. Surely, our brain never fails to surprise us!

Stress agents

A discussion of fear cannot be complete without at least a brief overview of stress hormones, which play an important role in the way our brain manages fear. Understanding stress could be relevant for unraveling the mechanisms underlying ethical behavior because stress affects fear circuitry: it can, for example, interfere with the suppression of fear, which, as my theory will show, can lead a person to behave according to the Golden Rule. Moreover, long-lasting stress, apart from affecting a person's general health, tends to impair memory, including the memory of fear, which, as I said at the beginning of this chapter, is vital for choosing to act ethically.

CRH

A master stress hormone involved in fear is a small Molotov cocktail of a molecule, the corticotropin-releasing hormone, or CRH, which exerts powerful effects on the brain and body. In outlining my theory in the next chapter, I am also going to argue that CRH could be one of the crucial molecules capable of triggering ethical behavior. The gene that produces this hormone is exquisitely responsive to stress, and CRH itself, in turn, causes a massive release of stress hormones.

For the purposes of our discussion it is important to know that CRH is produced not only in the hypothalamus, where its effects were first discovered, but also in the amygdala. Acting through its receptors in the amygdala itself and elsewhere in the forebrain, CRH stirs us to respond to a threat with fearful behavior. At the same time, it acts as the main trigger for the release of stress hormones in a chain reaction that runs from the hypothalamus to the pituitary to the adrenal gland. When this release is ramped up, steroidal stress hormones such as cortisol or, in a rat, corticosterone, pour out of the adrenal gland and help the entire body to deal with stress and fear.

Ned Kalin and Davidson have sleuthed out a link between CRH and innate fear. In the experiment I described in the preceding chapter, they saw that the innate, unlearned fears of monkeys were reduced when a portion of the animals' amygdala was destroyed. In that study, the University of Wisconsin researchers found that these monkeys turned out to have lower levels of CRH and less stress hormone pouring from the pituitary into the adrenal glands (hence the decrease in "freezing" behavior).

The CRH gene may also play a role in an important stress control mechanism that has become apparent from the use of cortisol or corticosterone as a treatment for, of all things, stress. In fact, from a genetic, or molecular-biological point of view, this treatment offers an opportunity to explore the most exciting and enduring effects of stress hormones in the brain. Alan Watts, from the University of Southern California, conceives the treatment's impact as a negative feedback loop that interferes with the functioning of certain genes. According to this concept, stress hormones turn off those genes in the brain that prod the brain into making more stress hormones, copying a natural negative feedback loop.

As suggested above, a prominent candidate for this feedback mechanism is the gene for CRH, because it causes the pituitary gland to release a hormone that, in turn, stimulates the release of stress hormones from the adrenal gland, which under normal conditions eventually sends back a negative signal that shuts down the CRH gene. When this natural feedback mechanism is not sufficiently effective, treatment with corticosterone turns off the all-important activity of the CRH gene. In fact, there is a clear-cut inverse relationship between the levels of circulating corticosterone and the activity levels of the CRH gene in the hypothalamus. The biological significance of such a simple negative feedback loop is that it holds the steroid stress hormones secreted from the adrenal glands at a constant level.

Watts is at pains to remind us, however, that the relations between corticosterone and CRH gene activity have complex features that go well beyond negative feedback. For one thing, the many different neurons with an active CRH gene respond in different ways to corticosterone. Not all cells respond with decreased gene activity—indeed, some, including neurons in the amygdala, increase it. That could influence our emotional state mightily after several stressful episodes. For another thing, taking off from the classic formulations of Mary Dallman at the University of California, San Francisco, Watts reminds us of variations in the dynamics of stress hormones over the course of time. Very fast actions, which dramatically reduce gene activity within minutes, differ from intermediate effects that might take two hours or more. The slowest kind of effect, taking place over the course of days of exposure to stress hormones, reveals the simplest negative feedback loop mentioned above: the inverse relationship between circulating corticosterone levels and activity levels of the CRH gene. Thus, research has uncovered a number of factors that could subtly influence the effects of stress hormone on genes and thereby exert an impact, eventually, on fear.

Vasopressin

Stress hormones released by the adrenal gland, such as corticosterone, also exert an influence on the activity of the gene that codes for the hormone called vasopressin: they powerfully inhibit the activity of the vasopressin gene in the hypothalamus. In fact, the activity of the vasopressin gene is at least as sensitive to circulating corticosterone as is CRH, and this sensitivity is important for our understanding of the molecular biological components of fear because vasopressin influences the efficacy of learning, including fear learning.

In evolutionary history, the hormone made by the vasopressin gene helped us conserve body fluid if we were wounded and losing blood. We will encounter this hormone a number of times in subsequent chapters, in connection with its roles in parental behavior and aggression. Its activity rate is extremely important because vasopressin can influence the arousal state of the central nervous system in four ways: by its hormonal importance, its effect on the autonomic nervous system, its actions in the lower brain stem to influence cortical electrical activity, and its direct behavioral effects.

Vasopressin's links to stress and fear have been confirmed by the studies of Rainer Landgraf, at the Max Planck Institute of Psychiatry in Munich, Germany. One reason Landgraf was able to breed rats for high-anxiety versus low-anxiety behavior was that the DNA "control switch" of the vasopressin gene carried a number of mutations. Landgraf reasoned that when the mutated versions of the genes were active, their abnormal control over the activity of the vasopressin gene would affect the rats' behavior: it would either contribute to the high-anxiety animals' excessive anxiety and fear—or, in low-anxiety rats, their abnormally low levels.

New agents of stress and fear

An additional player in the effects of stress hormones on the brain is the gene coding for the brain chemical enkephalin. This small molecule mimics some of the effects of opium and works in the brain through two sets of receptors to reduce nervous system arousal and the perception of pain. Some neurons producing CRH can also activate the gene for enkephalin. Stress-related biological variations in the body appear to affect the activity of the enkephalin gene in the same way they affect that of the CRH gene.

A recently discovered gene called M6a may help to explain the loss of synapses—connection between neurons—in the hippocampus of people with depression and in lab animals under experimental stress. This gene codes for a substance that helps to form the fibers extending from nerve cell bodies and connection-forming structures that stick out along the fibers; these structures are important because they dramatically increase the surface area, and hence the potential amount of signaling, of a neuron in the hippocampus, which, as mentioned above, plays a leading role in memory. According to Julieta Alfonso and her colleagues at Universidad Nacional de Gral in San Martin, Argentina, the M6a gene has a lower activity in animals that are undergoing physical or social stress. This and other studies suggest that certain hormonal mechanisms originating with CRH in the amygdala can relieve our stress or heighten our anxiety, depending on our emotional history.

New research continues to uncover fear-related genes that are active in the amygdala, although at first we often don't know exactly what these genes do. For example, recently a team led by Gleb Shumyatsky, at Rutgers University, issued a report on a mouse gene coding for a protein named stathmin, which has a noticeable effect on the personality of mice: it helps make them timid (more timid than normal, that is). Removing that gene by a set of biochemical reactions made the mice more daring: they failed to avoid innately frightening environments, and their fear memory was poor. Exactly how this happens is still unknown in cellular terms, but electrical recordings from cells in the amygdala show problems with the response to signals from the cortex and the thalamus.

Finally, a handful of stress- and fear-related molecules that affect the whole body may act differently depending on ways in which they are produced: the release of the adrenal cortical stimulating hormone from the pituitary gland follows several different patterns, including small pulses, a daily pattern, and a stress-dependent pattern. Genetic responses to stress hormones may depend on which of these patterns is holding sway. These complexities may one day tell us how chronic, repeated onslaughts of stress hormones influence our fears and anxieties.

Long-term stress, big-time effects

The effects of stress hormones on behavior are not simple; they depend on whether or not the levels of these hormones have been elevated chronically and repeatedly. If not, that is, if stress has been rare, their effect is restorative: it brings the body's systems from an emergency state back toward a normal state. However—and this point is crucial—if the animal or human has been subjected to chronic fear and stress, with adrenal hormones summoned up again and again, then the steroidal stress hormones can destroy neurons and impede signaling pathways.

Consider first the effects of acute, one-time exposure to stress hormones. Given before fear training, corticosterone can enhance the learning of fear in laboratory animals. Moreover, it increases the amount of CRH produced in the amygdala, running the fear system like a positive feedback loop: CRH stimulates corticosterone release, which in turn stimulates CRH, and so on. In other words, corticosterone increases the response to stress. Given after fear training, corticosterone can enhance the memory of fear (this is what caused the laboratory rats to stop all movement when the conditioned stimulus for an electric shock was sounded).

These phenomena likely depend on the amygdala: Jack Shepard and his colleagues at the University of Oklahoma showed that the placement of corticosterone directly into the amygdala makes the pituitary and adrenal glands churn out greater quantities of their hormones in response to stress. In another study, Yu-Ling Yang, Po-Kuan Chao, and Kwok-Tung Lu, from the National Taiwan Normal University and other universities in Taiwan, have reported that corticosterone in the rat amygdala helps to extinguish a learned fear response. (The latter result might seem like a contradiction of my earlier statement that the CRH can improve the learning of fear, but in fact it only reveals the complexity of the general picture: the CRH has an enhancing effect, which means that it can improve both the increase and the decrease of responses to stress.)

Stressors begin to interfere with memory when they become too severe or when they will not go away. Under some experimental circumstances, even a single injection of the animal stress hormone corticosterone can keep a lab animal from remembering to stay away from a stimulus that has frightened it before. Benno Roozendaal and McGaugh trained rats to inhibit a natural behavior in order to avoid receiving a slight electric shock to the foot. However, when the scientists injected the rats with corticosterone just before testing them, the rats were less able to remember their training and inhibit their behavior.

Robert Sapolsky, of Stanford University, has documented in laboratory animals the severe damage that can occur in the hippocampus, a part of the brain essential for memory, after excessive or prolonged doses of steroidal stress hormones. Since the hippocampus participates in shutting off hormonal stress responses, as explained by Bruce McEwen in his book *The End of Stress as We Know It,* a hormone that damages the hippocampus after prolonged stress makes things even worse. To paraphrase McEwen, stress hormones cripple the part of the brain that shuts them off, and so we have a vicious circle, which causes stress hormones to soar. By the way, some bad news: this vicious circle appears to strengthen with age. Worse news: it may even accelerate aging. These are very good reasons to avoid periods of prolonged stress, or, if they cannot be avoided, at least to break them up with relief intervals that will give the body systems a chance to recover.

Prolonged exposure to cortisol probably also holds great implications for human feelings and health because this hormone has been shown to affect emotions and memory in human studies. The laboratory of Richard Davidson at the University of Wisconsin has reported that research volunteers who received a single administration of cortisol subsequently rated a set of stimuli (which had been carefully selected as emotionally neutral) as more emotionally arousing than did volunteers who received a placebo (a dummy pill). Moreover, a single dose of cortisol strengthened the participants' memories of both emotionally loaded and neutral stimuli. These results suggest that a little bit of cortisol for a short time is good, but that a higher dose is harmful.

How can the positive, enhancing effects of an acute, onetime blast of stress hormones be reversed or even rendered harmful by chronic exposure? Let me count the ways (some of them first having been spelled out by Robert Sapolsky). First, Ron deKloet, then at Utrecht University in the Netherlands, figured out that there is not just one kind of stress hormone receptor in the brain but two: one of them is extremely sensitive to those first, short exposures to stress, but long and strong exposure brings in the second receptor, which activates an entirely different kind of molecular biological signaling system. For example, Louis Muglia and his team, at Washington University School of Medicine, have genetically engineered a mouse with a disrupted form of this second type of receptor in the forebrain; these mice, in which fear and anxiety were induced in the laboratory, responded to stress with abnormal kinds of movement. In short—two types of receptors, two types of reactions to the same hormones.

Second, repeatedly exposing the central nervous system to stress hormones can bring different neuronal systems into play, including those—such as the amygdala—whose long-term activation can mediate the deleterious effects of stress on fear memories. Third, as McEwen has shown, prolonged stress may ruin the extensions of neurons in the hippocampus, thus reducing memory capacity in general and fear memory in particular. Fourth, prolonged stress may slow the birthrate of new neurons in the hippocampus, a serious incursion, since those new neurons are thought to contribute to all our cognitive capacities. Fifth, stress renders even the neurons that are already present in the hippocampus more susceptible to deformation and death.

McEwen, writing in *The End of Stress as We Know It,* adds that many of the ways in which chronic stress affects our states of fear and anxiety may be *indirect;* they may not start in the brain at all. Chronic stress causes changes in the cardiovascular system, the immune system (lowering our resistance to illness), and the gastrointestinal system. Additional support for the long-lasting effects of stress hormones outside the brain comes from studies revealing that the adrenal glands, near the kidneys, can act independently of the brain. James Herman and his team, at the University of Cincinnati, have found experimental conditions in which the adrenal glands do not always act simply as slaves of the pituitary gland (the so-called master gland) and the hypothalamus. To give just one example, the researchers repeatedly exposed rats to the stress of perceiving the odor of a major predator: a cat. Although the rats showed an increase in gene activity for CRH in the hypothalamus, their adrenal glands did not become hyperactive in response, as might have been expected. In other studies from the Herman lab, experimental treatments did not bring on changes in the hypothalamus or pituitary gland but did incite the adrenal glands to produce extra corticosterone. Moreover, in an experiment that involved injecting CRH repeatedly into the amygdala of rats, Willie M. U. Daniels of the University of Stellenbosch and Dan J. Stein of the University of Cape Town, both in South Africa, have reported that the response of the pituitary gland to stress did not change, but that the adrenal glands began pouring out more corticosterone.

Thus, through all these causal routes, direct and indirect, the consequences of chronic stress—in other words, chronic exposure to high levels of cortisol or corticosterone—produce effects on states of fear that are quite opposite to those of a single, mild stress. I hope that in the near future, clinical research will give plenty of attention to the implications of these complex dynamics for the treatment of anxiety states, PTSD, obsessive-compulsive disorders, and other debilitating mental illnesses. Encouraging results have already been achieved in a number of studies. At Ludwig Maximilians University in Munich, Germany, Gustav Schelling and his team have reported reducing the intensity of symptoms of PTSD by giving cortisol, without affecting other aspects of memory. Likewise, at the University of Zurich, in Switzerland, Dominique de Quervain ran trials with PTSD patients in which even low doses of cortisol were able to reduce traumatic memories. However, in view of the dual effects of stress hormones on the brain, as documented above, I believe it will take a great deal of work to figure out which levels and schedules of hormone treatment are best for success for the individual patient in these therapeutic endeavors.

While effective therapies for stress- and fear-related disorders might be a thing of the future, we already know a great deal about the neuroscience of fear. My theory postulates that fear, learned and unlearned, comes into play in ways that help us to avoid acts of violence and other unethical practices. In presenting my theory further, I'd like to argue that the genes and neural circuits managing our fear supply crucial biological components leading to human behaviors that obey a universally accepted ethical principle, the Golden Rule.

5

LOSING ONESELF

What circuitry in the brain could possibly produce behavior as universal, as flexible, and yet as reliable as "fair play?" Believe it or not, according to my theory, this circuitry does not require fancy tricks of learning and memory but instead makes use of the easiest brain process of all—the blurring and forgetting of information.

From a scientific point of view, the best explanation of a complex phenomenon is the most parsimonious one, the hypothesis that accounts for all the relevant observations while resorting to the fewest, simplest steps possible. And scientists have an unexpected turn of phrase for such a hypothesis; they say it is "elegant." I hope to achieve such elegance by invoking the basic, primitive brain mechanisms that govern fear to explain Golden Rule behavior. Surely, I will not try to explain all of ethics through the neurobiology of fear, but mechanisms producing fear certainly are central to the explanation of how human beings behave according to one important principle. I will argue that the same relatively simple blurring process mentioned above—triggered by fear or other emotions—can account for a virtually endless repertoire of ethically acceptable behavior choices.

The theory has four steps. In the *first* step, one person considers taking a certain action with regard to another; for example, Ms. Abbott considers knifing Mr. Besser in the stomach. Before the action takes place it is represented in the prospective actor's brain, as every act must be. This act will have consequences for the other individual that the would-be actor can understand, foresee, and remember. *Second,* Ms. Abbott envisions the target of this action, Mr. Besser. *Third* comes the crucial step: she *blurs* the difference between the other person and herself. Instead of seeing the consequences of her act for Mr. Besser, with gruesome effects to his guts and blood, she loses *the mental and emotional difference* between his blood and guts and her own. The *fourth* step is the decision. Ms. Abbott is now less likely to attack Mr. Besser, because she shares his fear (or, more precisely, she shares in the fear he would experience if he knew what she was contemplating).

For the neuroscientist, this explanation of an ethical decision by the would-be knifer has one very attractive feature: it involves only the loss of information, not its effortful acquisition or storage. The learning of complex information and its storage in memory are deliberate, painstaking processes, but the *loss* of information seems to take place with no trouble at all. Damping any one of the many mechanisms involved in memory can explain the blurring of identity required by this theory. In the example of Ms. Abbott and Mr. Besser, as the result of a blurring of identity—a loss of individuality—the attacker temporarily puts herself in the other person's place. She avoids an unethical act because of shared fear.

Now that the theory is introduced, let's go through it step by step.

Step One. Represent my prospective action in my own central nervous system.

A long history of neurophysiological work tells us that our brain makes representations of actions we are about to take. In some cases, that scientific work also makes it clear why this is so. In the 1940s, the physiologist Erich von Holst reasoned as follows. If you are, say, taking a walk and admiring your neighbors' gardens, your angle of view on the tulips, the irises, and the flowering dogwoods must keep changing as you move your head and body. Is the world spinning around you? No. How do you know what is happening? Von Holst theorized that every movement is signaled to the individual's central nervous system. The information about movement is then compared to the resulting change in visual angle, so that, in our example, you can distinguish between the effect of your own movement and an actual change in the angle of the objects you see.

The same is true of hearing: when the loudness and the timing of sounds reaching our ears change in their balance between the left ear and the right, is that because we are moving or because the source of the sound is moving? The representation within our brain of the movements we make in the real world allows us to keep track. As for helping us keep our balance and a steady gaze when we move, that is the work of the vestibular system: while Mr. Audubon, for instance, watches a hawk's lethal plunge from mid-sky down toward a field mouse, the vestibular area (located near each ear) signals to his cerebral cortex that he is moving his head, so that he will not feel as if the ground has abruptly moved upward instead.

Von Holst grouped these ideas about representation of our muscular movements to our own brains under the name "reafferent theory." The scientist who carried out the most vigorous and successful tests of the implications of this theory was the engineer-turned-psychologist Richard Held. At Brandeis University and at the Massachusetts Institute of Technology, he performed experiments to study the need for active movements in our visual and auditory environments—movements duly represented in the brain—to maintain a sense of stability in our world. Most telling were experiments in which he had volunteers wear prisms over their eyes. The prisms changed the apparent visual angle in the volunteers' visual field by, say, 30 degrees. Therefore, if the volunteer reached for an object, her reach would be misdirected by 30 degrees. Only if the volunteer was allowed to make active movements while wearing the prisms, with those movements and their consequences relayed back to her brain, was she able to compensate for the prisms and reach toward the object accurately. The brain keeps constant track of our actions and their consequences, and these active movements and the signals emanating from them are crucial for normal eye-hand coordination.

In other experiments, infant monkeys were raised for 35 days with the ability to move their hands freely, but not to see them. At the end of that period, they were unable to make the accurate reaching movements they would normally have learned to carry out by then. Held concluded that active hand movements, along with the ability to see the results of those movements, were absolutely necessary for the regular development of visuomotor coordination. One of the mechanisms behind such coordination may already have been discovered: it is likely that certain neurons identified by Shintaro Funahashi and his colleagues at Yale Medical School, cells that manage the control of purposeful eye movements, must be brought into play for the development of normal eye-hand coordination.

A similar requirement for active movement holds true for at least one other mammal, the cat. Held and his team found that if kittens moved actively in their visual environment during development, their visual appreciation of distance and depth was normal. However, if they were simply towed around in a little gondola in that same visual environment, their depth perception was sorely inadequate. It seems that the ability to initiate and carry out a movement and to compare the reafferent signal of that movement with the visible results may be critical to a neurophysiological function such as depth perception.

The notion has cropped up that the brain keeps track of its actions in other parts of the nervous system as well. In a part of the forebrain essential for our sense of ourselves (the midline thalamus), the activities of neurons tell us clearly that we depend on premotor signals—that is, brain signals representing a movement that is about to happen—from our eyes and other parts of our bodies. A research group led by John Schlag, at UCLA Medical School, has extensively characterized these neurons in the thalamus of monkeys; the scientists found that such cells were active during eye movements and even that their firing patterns could anticipate changes in eye position. Schlag and Madeleine Schlag-Rey then discovered there were at least two kinds of cells: those responding to a visual stimulus transiently, to detect a stimulus toward which the eyes should move; and those with a sustained response, which provide a signal to hold the monkey's gaze upon the target.

Thus, as elaborated in the later results of Melanie Wyder and her colleagues at Wake Forest University, not all thalamic neurons keeping track of eye movements are the same. Their variations in the amplitude and exact timing of the electrical activity associated with rapid eye movements suggest a variety of ways in which this function—that is, keeping track of eye movements—is served. The thalamic neurons that perform this function are probably important for our ability to orient ourselves correctly toward a visual stimulus.

Nor is this type of self-signaling limited to higher animals and humans. Electrophysiological experiments with singing crickets, carried out recently by James Poulet and Berthold Hedwig at Ecole Polytechnique Fédérale de Lausanne in Switzerland, narrowed down the cellular basis for these copies, or representations, of our motor signals to a single type of neuron. In this simple animal, a powerful neuron that is both sensory and motor has the responsibility of coordinating the activities of several parts of the body and regulating the animal's ability to keep track of its own actions.

For another example of the same movement representation mechanism, consider a monkey looking at its reflection in a mirror.

Frans de Waal, at Emory University, asked whether the monkey can tell it is viewing itself rather than another monkey. The answer is yes: by comparing its own movements to those of the reflected image, it is able to recognize its reflection in the mirror as "special," according to several criteria such as more eye contact, friendly behavior, and less anxiety.

This ability to claim one's own reflected image is important not only for the recognition of self but also for the ability to understand and imitate others. In fact, according to electrical recordings from the monkey brain, certain parts of the cerebral cortex contain neurons that show a high degree of electrical activity not only when the monkey performs a particular action but also when the monkey observes another monkey doing that same action. These specialized cells are known as "mirror neurons." As described by Giacomo Rizzolatti, at the University of Parma, Italy, the activities of such neurons in the human brain are involved not only in recognizing another's actions but also in understanding other people's intentions.

"Intentions"—can cells really recognize intention? If so, these cells sound like an extremely sophisticated subject for analysis, even for modern neuroscientists. But Rizzolatti and his colleagues do not rest there; they make a further claim that our conscious intentions also have a neurophysiological reality. The Parma research team has observed that within special parts of our cerebral cortex, electrical activity is higher if we think about our own intentions. In Richard Andersen's laboratory at the California Institute of Technology, nerve cells in a part of the cortex called the "parietal reach region" showed electrical activity that predicted how, when, and where the hand was about to reach. To quote Marco Iacoboni, at the University of California, Los Angeles, "You understand my action because you have in your brain a template for that action based on your own movements." This is how "mirror neurons" could be involved in the empathic understanding of another person's actions.

These new and very exciting studies of mirror neurons fit into a long neuroscientific history of study about the representation of movement in our brains. In the 1960s and 1970s, Edward Evarts, at the National Institutes of Health, recorded individual cortical neurons to show their activity before the movement occurred. Poring over brain-wave recordings, Herbert Vaughan and Hubert Kornhuber at the University of Tübingen, Germany, detected neurophysiological signs of "movement intention" that they called "readiness potentials." There is nothing mysterious about the brain's ability to represent movements to itself. Even a computer-controlled robot can be programmed to plan motor controls in a predictive manner that produces more sophisticated movements than simple reflexes.

In sum, a brain must keep a running record of what it is doing in order to monitor the relations between an action and the consequences of that action. That is how it maintains stability in a changing world. Although my theory of our capacity for ethical behavior depends on this ability, in no way are these neural mechanisms limited to ethics, nor are they "special" for ethics. Neurophysiologically speaking, this is old news.

Step Two. Envision the target of this action.

How, in our mind's eye, do we envision the person toward whom we might act? Figuring this out is no small task. Even the great advances in the neuroscience of vision during the second half of the twentieth century dealt only with extremely simple visual stimuli, such as individual lines at particular angles. These discoveries were made by recording from individual neurons in the primary visual cortex, the part of our cerebral cortex that receives visual signals directly from the thalamus. Meanwhile, we knew that some mechanisms specialized for recognizing faces would crop up sooner or later, because the great neurologists of the late nineteenth century had found that people who suffered damage to the cerebral cortex (for example, from injury or stroke) had trouble afterward with perceiving human faces *as* faces.

In the early 1970s, Charles Gross, working at Princeton University, began to find startling properties of cells in a part of the cortex he had been studying for a long time, the inferotemporal cortex—an area that was far removed from the primary visual cortex but that nevertheless responded to complex visual stimuli. Most important for our purposes, he found inferotemporal cortical neurons that responded *only* to faces. Some neurons responded best to faces shown in profile, others to frontal views. Although these peculiar electrical recordings were obtained from nerve cells in the monkey brain, recent work at MIT has shown that similar mechanisms must operate in the human brain. In the Department of Brain and Cognitive Sciences, Nancy Kanwisher and her team have found a convincing "face recognition module" in the inferotemporal cortex of the human brain, located just behind and below the ears. She has even identified specific electromagnetic waves associated with the categorization of visual stimuli as faces and the successful recognition of individual faces.

For a long time, theoretical neuroscientists had supposed that our ability to picture other folks in our minds would depend on what were known as "grandmother neurons"—the ultimate in specialized cells, one assigned to each person of our acquaintance. Each neuron would fire if and only if the image of its own particular person, for instance, our grandmother, passed in front of our eyes and therefore allowed us to recognize our grandmother. However, Charles Gross thinks that each face-selective neuron is actually a member of numerous ensembles of neurons for coding faces. Indeed, Keiji Tanaka, in the Riken Brain Science Institute in Saitama, Japan, has found evidence of overlapping fields of neuronal activation by similar facial stimuli. Yasushi Miyashita, in the Department of Physiology at the University of Tokyo, thinks of different inferotemporal cortex neurons as encoding different "letters in a perceptual alphabet." For example, individual cells might respond only to particular features of the face or head. As a result, facial recogni-

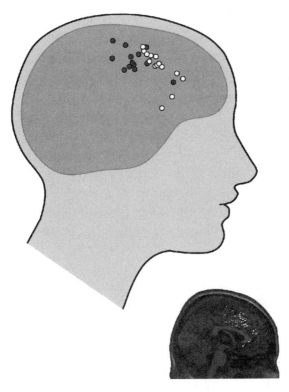

Figure 6. Feeling your pain. Many brain areas participate in the experience of pain. Among them, the anterior cingulate cortex and the insula help give pain its emotional quality. Studies by Phillip Jackson at Université Laval in Quebec and his colleagues have found activity in these two circuits not only in response to one's own pain, but also in perceiving someone else's pain. The dark circles represent brain responses to one's own pain and the white circles locate responses to perceiving another's pain. *Photo Image and mapping from Fig. 1(A) in PAIN 125 (2006) 5–9. Used with permission from IASP₀.*

tion depends on patterns of activity across populations of neurons in the inferotemporal cortex. This is a more complex and ultimately more satisfying explanation than "grandmother neurons."

Thanks to the intricacy of these patterns, humans have an enormous capacity to recognize faces (some of us more than others!), but this capacity is also easily reduced. If any one part of the multistaged set of mechanisms by which visual information reaches the primary visual cortex and then gets transmitted to the inferotemporal cortex should happen to misfire, the failure would destroy our ability to form the mental image of our intended actions. Faces are recognized by groups of neurons that the late, great Canadian brain scientist Donald Hebb called "cell assemblies." In the infero-temporal cortex, if any one part of a cell assembly were not working properly, it would reduce our ability to recognize and think about a person's image.

Recognition is partly supported by the same neuronal circuitry as our sense of self, and this circuitry is extensive and versatile. Even in the human brain, so dominated by visual systems, the sense of self need not be limited to envisioning oneself. For Antonio Dama-sio of the University of Iowa, author of *Descartes' Error* among many other publications, a sense of self depends on "numerous brain systems that must be in full swing." These include "representations of our musculoskeletal frame and its potential movement," reconstructed on a moment-by-moment basis. Several recent studies have pictured wide-ranging circuits in the human cerebral cortex that are activated by recognizing one's own face as compared with others. The sensation of being touched on one's own person seems to work in much the same way. For my purpose in this book, the most important implication of these multiple and complex circuits is their susceptibility to blurring and confusion. Any part of them, damped down, would allow us to run our sense of self together with our sense of another human being—indeed, this is precisely what I hypothesize happens next.

Step Three. Blur the identity of the target with my own.

This is the easy part. Think of any and all complex devices: is it easier to change them from not-working to working or from working to not-working? Obviously, the latter. It is difficult to construct complex devices so that they work, but it is very simple to interfere with their operation or unplug them. Consider a computer, or even a remote control device for the television: as most of us know all too well, even small changes in the hookup or operation of these gadgets can reduce their performance dramatically.

Therefore, it is clearly plausible for me theoretically to propose a reduction in the operational efficiency of neural circuits that discriminate between self and not-self. Reduced performance by *any* of the sophisticated cerebral cortical circuits and mechanisms sketched in step two would blur individual identity.

Where does the reduced image processing take place? In the cerebral cortex. How does it take place? I think the blurring of identity can occur in two easy ways that are not mutually exclusive. In both neuroanatomical explanations, cell assemblies responsible for encoding information about individual identification come into contact with each other, either in their original cortical region or elsewhere, where an "ethical switch" is thrown.

Blurring through excitation

First, within an individual cortical region responsible for generating images of people, an overall increase of excitation could cause the neurons responsible for signaling the identity of one person to signal the identity of another as well. Neuroscientists have already found three different cellular mechanisms that could cause such overexcitation: the phenomenon known as "disinhibition," increased excitability from the neurotransmitter acetylcholine, and gap junctions. These, too, are not mutually exclusive; they all could work.

"Disinhibition" refers to the loss of inhibition that is normally achieved by the inhibitory transmitter GABA at the synapses, the tiny spaces at which neurons communicate with one another. Inhibitory relations between neurons in the cerebral cortex are delicate and changeable. For neurons signaling visual stimuli, for example, even a small change in the spiking frequency of inhibitory interneurons is important. Any impact from inhibitory neurons on the cells that signal for visual stimuli could be brought about simply by the phosphorylation of one subunit of a single GABA receptor—that is, the addition of a phosphate group, which changes both the shape and the electrical charge of the protein. A chemical change of this nature would reduce the receptor's efficiency enough to cause disinhibition. Theoretical neuroscience tells us that mutually inhibitory relations among neighboring neurons and the overall balance between excitation and inhibition tend to display dynamics that can account for a wide range of behavioral results, even when the basic connectivity among these neurons remains the same.

A second detailed cellular mechanism would entail increased excitation from the neurotransmitter of arousal, acetylcholine. The effects of acetylcholine on the cortex are not simple. David Prince and his colleagues at Stanford University Medical School used electrical recording from neurons in the visual cortex to look at the effects of acetylcholine on two different kinds of inhibitory interneurons, called "low-threshold spike" cells and "fast-spiking" cells. While acetylcholine excited low-threshold spike cells through one type of receptor, it inhibited fast-spiking cells through a different type. John Hablitz's lab at the University of Alabama has used a voltage-sensitive dye to discover that acetylcholine-related influences enhance synchronized activity among these interneurons; this synchronization, in turn, propagates waves of activity in the cortex.

The importance of these neurophysiological results for my theory can be appreciated with just a moment's reflection on how acetylcholine works, which is in a nicotine-like manner. Nicotine clearly enhances cognitive function and is an important influence on the overall arousal state of the cerebral cortex. In similar fashion, acetylcholine is known to be able to reorganize the maps of sensory information in the cortex. Therefore, fine differences in the amount of acetylcholine released in the cortex could greatly affect the spread of excitation, depending on the exact location of its release. A considerable spread of excitation could even merge the images of two distinct people—say, the image of would-be assailant Ms. Abbott with that of her intended victim, Mr. Besser.

A third possible mechanism for the direct merging of two cell assemblies representing individual identifications could also involve the rapid spread of excitation within a particular region of cerebral cortex, but not because of acetylcholine. Rather, excitation could spread because of the operation of gap junctions, infinitesimal spaces between nerve cells that allow a direct spread of electrical current from one nerve cell to another. Gap junctions are a natural marvel of engineering: proteins bound together in the form of tiny barrels, with six "staves" in each of the two cells. The interface of the two barrels in the two neighboring neurons is what creates the gap junction. Importantly, it is very much easier for one cell to excite another through a gap junction than it is through a synapse. Whereas a synapse is relatively uncertain and slow (requiring several milliseconds to complete), a gap junction is certain and fast (taking a fraction of a millisecond).

Nancy Kanwisher of MIT, who is an expert in face recognition by modules of neurons in the cerebral cortex, points out how easy it can be for these cellular mechanisms of blurring to operate effectively in degrading individual recognition. She mentions that simply adding points of "noise" to an image, or even throwing off the exact time of arrival of various components of the image, would significantly reduce individual identification.

For all three of these detailed cellular mechanisms, we need not be thinking of a wholesale junction of, for example, one facial image into another. Because an image is composed of many lines, angles, and curves, the effective blurring of one image into another could be achieved simply by achieving more similar activity from line to line, angle to angle, and curve to curve. All three mechanisms provide for a merging of the image of the target of one's action with the image of oneself, the actor.

Blurring through emotion

Another neuroanatomical approach to mechanisms for identity-blurring provides still another plausible, specific theory. This is not inconsistent with the first approach; in fact, the two could operate in parallel. This approach takes place outside the cortical region in which individual images are processed. Visual signals that identify individuals—and for that matter, signals from other sensory modalities such as sound, touch, and odor—are sent from the visual cortex and the inferotemporal cortex to another part of the forebrain. Which part? Two possible places at which these images might converge are within the amygdala itself (which we have gotten to know as a "headquarters" of fear), because of its control over emotion, and within the prefrontal cortex, because of its suppression of certain forms of emotional behavior.

I like this neuroanatomical theory as much as the first. There is good precedent for sensory information connected with moral judgments being sent to areas of the brain traditionally connected with emotion. In 2003, an American and Italian team of scientists used functional magnetic resonance imaging (fMRI) to look at neuroanatomical patterns of brain activation in volunteers who were either imitating, or simply observing, emotionally loaded facial expressions. Laurie Carr and Gian Luigi Lenzi found greater activity during imitation than during observation, not only in several sites below the cerebral cortex (for example, the hypothalamus and the septum, whose activation produces involuntary reactions such as rage, fear, tears, or orgasm) but also in a primitive part of the cortex itself. According to the authors, sending the sensory/motor information to an area connected with emotion allowed the person to experience a feeling of empathy.

Some studies have used brain imaging to see volunteers' responses to pain they feel personally versus pain they perceive being experienced by others. As described by Philip L. Jackson and colleagues in a 2006 review of this research, some of the same brain circuits for pain become activated whether the person is receiving the painful stimulus or observing another person receiving it—but the activations differ in intensity and occur in slightly different locations within the circuitry.

Approaching the topic from a different angle, Jon Cohen and his coworkers at Princeton University asked volunteers to consider each of a battery of 60 practical dilemmas. The 60 were divided into moral dilemmas that had a strong personal component, moral dilemmas that did not have a personal component, and nonmoral problems. The scientists measured the effect of exposure to these problems on brain activity, using fMRI. The results were clear: moral dilemmas with personal content not only activated parts of the cerebral cortex that have been associated with emotional processing, but they did so to a much greater extent than moral dilemmas that had no personal impact. The activation was greater for those problems, in turn, than for nonmoral problems. Thus, the Princeton experiment gave evidence for the flow of events in my earlier example when my would-be stabber decided not to act: the passage of information from purely sensory or motor parts of the brain to another part of the brain, in this case with emotional import, where a moral decision can be made.

Even the brain of a rat has the capacity to transfer purely sensory information to an emotionally and ethically important brain region such as the amygdala. Ewelina Knapska and Tomasz Werka, at the Nencki Institute in Warsaw, Poland, carried out a fascinating experiment on about a dozen pairs of rats. Within each pair, one rat, the "demonstrator," was exposed either to a slight electric shock to the foot or to a neutral but novel environment. What would be the influence of the demonstrators' experimentally imposed different states of fear on the exploratory behavior of the other member of the pair, the "observer?"

It turns out that the emotionally laden experience of the demonstrators who received the foot shock did affect the exploratory behavior of the observers. Those rats who observed the behavior of foot-shocked demonstrators explored their environment significantly more than those who observed the behavior of nonshocked demonstrators. (This reaction may seem counterintuitive: if anything, one might expect individuals that saw their cohorts receive an electric shock would explore *less*, if indeed they did not freeze in place altogether. Possibly the observers explored more because they were looking for a way to escape from a perceived threat.) But the most striking observation from this experiment is that even though these animals had not undergone an uncomfortable experience themselves, they recognized signs of discomfort in the other animals' behavior, and changed their own behavior accordingly.

And the response of the observers was not limited to their behavior; there were also differences in the activation of cells in specific parts of their amygdala. In the basal and lateral cell groups within the amygdala, the observers who saw the behavior of the shocked rats had greater neuronal activation than those who saw the behavior of nonshocked rats. In higher animals, sensory information of this sort would be connected with empathy; even in rats, data had been transferred to the amygdala.

Activating an ethical mechanism

Thus, an ethical "switch" could reside, for example, in the prefrontal cortex or the amygdala. When the switch is flipped into one position, it "turns off" the self-other distinction and the images of two individuals—say, myself and the target of my intended action—are merged. The result is empathic behavior that obeys the Golden Rule. When the switch is flipped into the opposite position, the images remain distinct from each other, and I may harm the target. Notably, even though I have usually described these images as visual, none of the theoretical steps I have brought up is limited to one sensory modality. Sensory stimuli having to do with hearing, movement, touch, and smell could all figure in.

What could control such a switch? Clearly, the prime candidates are substances that have already proven to be important for controlling the social behaviors of higher animals and of people toward one another. In particular, the gene that codes for the hormone oxytocin and its receptor, which we will meet in the next two chapters in connection with sociability, should be considered for this role. My colleagues and I at Rockefeller University are in the process of identifying other brain chemicals that could do the same thing. For example, vasopressin, another hormone, could be important in this regard.

Another candidate to control the switch is the corticotropin-releasing hormone, CRH, the master stress hormone I introduced in the preceding chapter. It could be involved because of its importance in fear and anxiety, and it might account for feelings of fear involved in ethical behavior.

Finally, all three of the detailed cellular mechanisms described above in connection with the theory presented in this chapter could be important here, too. Thus, even at this early stage of theorizing, we have plenty of realistic cellular and molecular candidates to do the job of making an ethical decision.

This neuroanatomical theory for the merging of images proposes the passage of sensory information relevant for a moral decision from sensory/motor parts of the brain—or, in neurophysiological slang, the "Apollonian nervous system"—to parts of the brain connected with emotions—the "Dionysian nervous system." In fact, some theoretical neuroscientists would argue that to break out of purely sensory and motor processing and into the area of ethics, exactly this sort of transition is required. They might argue from the neuroanatomical divisions of the brain charted by the great neuroanatomist, Paul MacLean, of NIMH. MacLean conceived of the brain as divided into three huge parts: (i) the most recently developed cerebral cortex, typical of higher mammals; (ii) the more primitive forebrain, connected with emotions and drives; and (iii) the most primitive vertebrate brain, required for maintaining vegetative functions. Such a theorist would say that I have proposed a flow of ethically relevant information from (i) to (ii) for the purpose of operating the ethical switch discussed above.

Loss of information

The loss of information—an individual's blurring of the distinction between himself and the intended target of his action—constitutes an essential element of my theory. The recognition of another person depends on a long series of electrophysiological and biochemical reactions to the stimuli particular to that person. These include not only seeing the other person's face, hearing his voice, feeling his touch, and smelling his personal odors. Every one of our senses must work hard to identify that someone is that very person, there, and not another person and not ourselves. Reduction of the ability to make such discriminations in any of these sensory pathways will result in a blurring of the target's identity. His identity becomes less easy to discriminate from others and, in fact, from oneself.

Explaining sensory discrimination and the processing of sensory information is a daunting task, even trickier than explaining how a computer works. But the *loss* of information is easy to understand. Therefore, it is easy for us to see how one could associate the consequences of an unpleasant act with oneself. Consider the many mechanisms that I have argued would provide that essential blurring of identity. The fact that there are so many different ways to accomplish the blurring of identity makes it especially likely that the overall theory actually works in the manner I have proposed. They all could work independently to foster identity-blurring, and in fact different ones could be most important in different individuals. But, for everyone, joining the other person's identity to your own will make it easier to empathize with her or him.

Step Four. Decide.

If the consequences of my action are potentially harmful for the other individual, I will probably not take that action, because I would not want to be harmed myself. If it is positive and helpful for the individual, I am likely to do it, because I would like to be pleased myself. The ethical decision depends on the blurring of identity between myself and the target of my intended action. Because of the blurring, we put ourselves in the other person's place. If that person would be afraid, so will we sense fear. Shared fates, shared fears.

These four clearly demarcated steps, with scientific substantiation, link the neuroscience of fear to the performance of an ethical act. Most of the situations and brain activity I've brought up so far have emphasized the negative—unethical acts and fear of their consequences—but that lopsided picture is just half the story. Fortunately, a whole symphony of other brain processes point us toward following the Golden Rule in an affirmative way as well. We are well supplied with brain mechanisms for pleasant, loving, prosocial acts, and these, too, support ethical behaviors.

6

SEX AND PARENTAL LOVE

"If our two loves be one." "Two hearts beat as one." "Tonight is the night when two become one." From John Donne to U2 and the Spice Girls, our culture is replete with expressions of yearning for a union of two beings. The positive version of the Golden Rule, one that impels us to love other human beings as ourselves or at least to treat them well, points in the direction of such a union of souls. Having established how fear leads to the blurring of identity that prevents us from doing harm to others, let us now turn to the positive reasons for erasing the borders between "me" and "you," or "self" and "other": blurring for love.

While the brain mechanisms of fear and love are quite different, the conceptual framework connecting both to ethical behavior is the same, providing for an efficiency that, from a scientific point of view, is most pleasing. The same identity-blurring principles I have outlined in the preceding chapter as arising from shared fear can also be triggered by positive and friendly shared emotions. It is no accident that two terms describing crucial components of friendly behavior, *compassion* and *sympathy*, are derived from the Latin and Greek words meaning "feeling with"—just the kind of a shared sensation needed for turning on the "ethical switch" that (usually) prompts us to behave in a prosocial, considerate manner.

In evolutionary terms, it all began with sex: One can hardly imagine a more definitive merging of two individuals than the one that occurs in the sexual act. What a powerful incentive for softening the "self-other" distinction! Later on in evolutionary development, the primitive mechanisms supporting sexual behavior gave rise to more complex interactions of a positive, cooperative sort— first and foremost among these the love of a mother or father toward their children. Sexual and parental love, in turn, set the stage for the underlying mechanisms of all other friendly and social behaviors. Thus, tracing the neuroscience of sex and parenting, as we will do in this chapter, could reveal how the biological roots of the Golden Rule might have evolved.

Mating

Biologists talk about sexual relations between males and females as the basis, or *bauplan* (a German term meaning "fundamental structure"), for all social behaviors. Moreover, despite the emotional complexity attached to it by humans, in biological terms sex is governed by some of the most basic and simple brain mechanisms compared with other endeavors, and neuroscientists turn to it as one of the easiest behaviors to study. I'd like to argue that sexual relations not only help us discover how the brain controls behavior in general, but also, more specifically, they help us understand how the friendly social behaviors might work.

To appreciate the central role of sexual behavior for the evolution of all behaviors—including the most complicated forms of human interactions in modern society—consider that sex is at the leading edge of evolutionary change. The prominent twentieth-century evolutionary geneticist Ernst Mayr of Harvard University proclaimed that "almost invariably, a change in behavior is the crucial factor initiating evolutionary innovation." Evolutionary thinkers speak of natural selection as a key step in the development of new characteristics of any species. In behavioral biology, therefore, the so-called selection pressure is greatest for the biological steps connected most intimately with the passing on of genetic material to the next generation. Among behaviors, those biological steps would be the ones related to sexual activity, because fertilization results in combining two hereditary histories, of the male and the female. To quote Mayr, sexual reproduction provides "the inexhaustible source of every new combination of individual variations such as are indispensable for the selection process."

In most mammals, the female controls the rate of progress toward mating behavior. This makes biological sense, because a good part of her energy for a long time will be devoted to pregnancy and then to maternal care. Let us therefore consider the hormonal mechanisms underlying female sexual behavior in some detail. As the eggs develop inside a female's ovaries, these eggs secrete estrogens—flat, rigid steroid hormones that circulate in the blood and eventually flood the entire brain. In my research I have found nerve cells in some regions of the brain to contain proteins that fold in an unusual way. These proteins are estrogen receptors; they form a pit that takes in estrogens and retains them in the nerve cell for hours, so that the hormones can affect behavior for quite a while. Importantly, a significant number of these estrogen receptor–containing neurons are located in regions of the forebrain that play prominent roles in sexual feelings and mating behaviors.

How do the estrogens influence the activity of the neurons and, consequently, animals' behavior? In two ways: rapid and slow. In

SEX AND PARENTAL LOVE

their rapid action, estrogens kick-start the nerve cell as soon as they come into contact with its cell membrane. In the slow mode of action, the estrogens, once coupled to their receptor proteins, gradually work their way into the nucleus of the neuron. There they attach to DNA, thereby turning on or off an entire panoply of different types of genes. Some of the genes activated by estrogens produce substances that alter activity in the neurons of the hypothalamus and the ancient forebrain, which control mating behaviors in lower animals and affect sexual feelings and behaviors in humans. In lower animals, the female's mating behaviors are restricted to the time of ovulation. In humans, erotic feelings and sexual behaviors clearly are not so restricted, but several reports from evolutionary psychologists claim that promiscuous behaviors by women are more frequent at midcycle, when sexual activity is most likely to result in pregnancy.

What's relevant from the point of view of our discussion is that some of the substances whose release is triggered by estrogens are involved not only in sexual behavior but also in social recognition and other friendly behaviors. Among these are a brain chemical called enkephalin and its receptor, which inhibit pain under a wide variety of circumstances. I believe that these brain chemicals, by causing a partial painkilling effect, permit females to put up with mating behaviors by males that might otherwise be obnoxious. Even more relevant to sociable behavior is the hormone oxytocin, which, as we will see later, appears to contribute to the blurring of the "me-you" distinction in a variety of social situations.

As for males, since they abide by the Golden Rule no less than do females (though some women might disagree with that), it is just as important to understand the mechanisms underlying their courtship and sex behaviors. These mechanisms are similar to those of the female in many ways. Male sex hormones such as testosterone are secreted by the testes, circulate in the blood, and bind to their appropriate receptors in nerve cells in the brain. There, in different nerve cell groups, they exert their action in a rapid or slow manner, much like the female sex hormones described above. Testosterone-driven changes in the forebrain activate neuronal patterns necessary for erections and ejaculations, and they also activate courtship behaviors toward the female that will result in mating.

While searching for the brain mechanisms of ethical behavior through a study of mating, we must keep in mind that the situation is somewhat complicated by the difference in behavior between males and females. Not only are their sex hormones different, but their brains respond to the same hormone treatment in a different way. When we examine patterns of gene activity in the brain after such treatments in male and female laboratory animals, it is easy to see, at least in lower animals, how sex differences in behavior come about. For example, estrogen treatment in female laboratory animals turns on certain genes involved in sexual behavior in those parts of the female brain that support her body's preparations for reproduction; when given the same treatment, the male fails to generate the same chemical in his brain. In fact, the same gene—for example, a gene coding for estrogen—can have different behavioral effects, depending on whether that gene is being activated in a female or in a male. (The males have the same estrogen receptors as the females, but these receptors work differently, possibly because the males lack certain molecules called "co-activators," that in females affect estrogen activity.)

The multitude of genes involved in the control of sexual behavior of female and male mammals stands in sharp contrast with the relatively simple state of affairs revealed in studies of much lower animals like fruit flies. In the fruit fly, according to the geneticist Barry Dickson at the Institute of Molecular Biotechnology in Vienna, Austria, a single gene is sufficient to determine all aspects of a fly's sexual orientation and behavior. I suspect that, in humans, the very complexity of our genetic and neural controls over sexual behaviors has facilitated the development of mechanisms supporting social behaviors far more subtle than copulation. Not only must we adults live long enough to reproduce and tend to our young for a decade or two but we also need to exhibit some characteristics that are sufficiently appealing to induce others to mate with us. Among humans, a number of the same social, intellectual, and emotional traits that render us attractive as potential mates also support positive, cooperative behaviors involved in the ethical treatment of other people.

In 2006, thoughts about sexual behavior and selection of the sort I espouse became susceptible to mathematical treatment. Joan Roughgarden and her colleagues at Stanford University's Department of Biology published a paper proposing to incorporate theories of sexual attractiveness into a branch of mathematics called "cooperative game theory." According to Roughgarden, straightforward equations can give voice to a form of bargaining among individuals in the social game. These individuals could be, for example, a pair of potential mates who will provide various benefits to each other. Importantly, however, for my present thinking is that they don't have to be males and females engaging in courtship. Rather, the mathematical game supported my idea that the same cooperative principles governing sexual bonding can be extended to apply to a wide variety of social relations in general—and parenting in particular.

Motherly love

If there ever was a symbol of love that knows no limits, it is that of motherly love. In evolutionary terms, it has appeared later than sexual attraction and has taken the blurring of identity between two living beings to new heights, fostering robust brain mechanisms of care and compassion. Along with fatherly love, it has provided the evolutionary basis for social bonding of all kinds and the various forms of supportive, ethical behavior of people toward one another.

Maternal care requires so much time and energy that it should not be taken for granted. In the laboratory, for example, a mother rat with a full litter of babies may spend as much as one-third of her body fluid per day supplying milk to those babies. Obviously, when a mother takes good care of her children, it must happen for a reason. Susan Allport, writing in her *Natural History of Parenting*, conceives the answer as a simple matter of efficiency. Is it a better use of animals' resources to produce more and more eggs and sperm, show more and more courtship and sexual behaviors in order to copulate and eventually to forget about the young, or is it more efficient to take better care of the babies they have produced? In higher species, the efficiency certainly favors parental care that is more thorough and continues for a longer period. Allport feels it has evolved especial-ly well in predictable environments, in which parents would develop successful strategies for the survival of their offspring, or in environ-ments in which the young are most susceptible to predators.

What is maternal behavior all about? Since most mammalian ba-bies cannot care for themselves, they must be brought to a place where they can be kept safe, warm, clean, and adequately fed. For instance, in a large litter of laboratory mice or rats comprising nine or ten pups, this means that the mother will find the pups that are scattered about, carefully pick them up in her mouth, bring them back to a nest she has constructed to hide and keep them warm, crouch over them, and feed them with her milk. In fact, when we use the term "mammals," we are talking about species in which the females have mammary glands for feeding their young. As the young mature and become competent to move about on their own, breast-feeding and other maternal behaviors correspondingly de-cline. In humans, however, the care and compassion associated with motherhood usually persist throughout a woman's lifetime.

The hormonal tides that flush through the female's body before and during motherhood pertain to our discussion because along with the purely physiological effects on various organs, they affect maternal behavior. Perhaps the best-known hormonal determinants of maternal care in laboratory animals have been the estrogens, which circulate in the bloodstream at very high levels toward the end of pregnancy, and progesterone, which is produced at high levels throughout most of gestation and, for unknown reasons, drops shortly before delivery. Neuroendocrinologists have found that maternal behaviors in these animals can be best encouraged by a sex-hormone treatment mimicking this natural pattern: high doses of a form of estrogen called estradiol, accompanied by high doses of progesterone that are discontinued after some time.

Interestingly, progesterone produces its effects on behavior not only in its original chemical form; it is also transformed in the body into other, slightly different versions that can have a tranquilizing effect. Since these progesterone versions are dominant during nursing and we know that disturbing a nursing mother can interfere with maternal behaviors, it could be that these hormones are important simply because they help the mother to stay calm.

Less well known than estrogens for their involvement in maternal behavior are an assortment of hormones that include prolactin, the prostaglandins, and even a digestive hormone called cholecystokinin. Their effects on behavior (and in humans probably also on the woman's maternal feelings) vary in magnitude, and they may be acting directly, by affecting the brain's neurons, or indirectly, by stimulating the release of other, behavior-altering chemicals.

As its name suggests, prolactin, produced by the placenta and the pituitary gland, is necessary for normal lactation. It also fosters normal maternal behavior, as we have learned from animal experiments, where it appears to support the animal's tendency to retrieve her pups in a well-organized, careful manner. It turns out, however, that prolactin release from the pituitary gland can be influenced by the mother's psychological state, a fact that can be deduced from the finding that the release of this hormone is affected by an assortment of brain chemicals and transmitters that vary with changes in mood. I always had assumed that prolactin produced outside the brain somehow circumvented the blood-brain barrier, the thick layer of cells preventing most blood chemicals from entering the brain, and in that way affected behavior. I was surprised, therefore, when Richard Harlan and Brenda Shivers, in my lab, found that the prolactin gene is active in nerve cells in the hypothalamus, meaning that prolactin can be produced in the brain itself. Extensions of these nerve cells spread not only to the preoptic areas, the parts of the cortex important for maternal behavior, but also to the brain stem. Phyllis Wise and her team, then at the University of Kentucky, as well as several other groups, similarly found widespread activity of the gene coding for the prolactin receptor in many areas of the brain. Prolactin receptors, large proteins in the cell membranes, bind to prolactin and trigger a series of chemical events in the cell's cytoplasm. The loss of the prolactin receptor gene, which has not yet been studied in males, significantly harms maternal behaviors in the female mouse. All these findings, revealing the widespread activity of prolactin and its receptors in various parts of the brain, lead me to infer that the broad impact of this maternally related gene system could support not only maternal behavior but also a range of nurturing behaviors toward other members of the same species.

Prolactin is just one member, albeit an important one, of a family of large proteins connected with lactation. Certain other proteins in this family, known as placental lactogens, also foster maternal behavior under experimental circumstances, according to the results of Robert Bridges, at Tufts University. The proteins are synthesized in the placenta and are essential to pregnancy. To my way of thinking, the maternal behavior–stimulating effects of prolactin and placental lactogens are intended to bring the animals' behaviors into synchrony with the biological realities of their bodies.

The prostaglandins compose yet another set of hormones through which maternal behavior is brought into sync with the body as the mother prepares for delivery and gives birth to her young. Not only do the prostaglandins hold major importance for the contraction of smooth muscles, such as those lining the uterus, but one specific prostaglandin, F2alpha, has been shown to rapidly

Successful Maternal Behavior

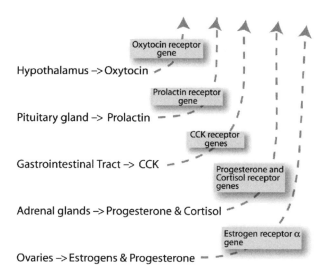

Figure 7. A fundamental complex. Several genes and hormones work together to support maternal behavior

stimulate the display of normal maternal behaviors in some animals.

Cholecystokinin, though known primarily as a protein fragment released from the gastrointestinal tract during feeding, plays a bit part in our story: when injected into female laboratory animals in the presence of newborns, it rapidly facilitates maternal behavior. It may be acting directly through appropriate receptors in the brain, but its effect might also be indirect, as it stimulates the release of prolactin and of the leading hormone of parental behavior, oxytocin, which will receive special treatment in the next chapter as a prime candidate for upholding the Golden Rule. Thus, a variety of genes and hormones essential to pregnancy also help to turn on behaviors required for nurturing the resulting babies.

It must be noted that apart from nurturing, maternal care sometimes requires a high level of aggression to defend the young. However, maternal aggression appears to be tightly controlled by specific genes that play no role in other forms of aggression. Hence, eliminating different aggression genes in laboratory-animal mothers can produce very different results. For example, Sonoko Ogawa in my lab at Rockefeller has found that female mice lacking a copy of the gene coding for estrogen receptor "beta" show higher levels of maternal aggression—that is, combativeness in the face of a threat to their offspring—but these genetically altered mice shared lower levels of aggression brought on by the injection of testosterone, compared with sister females who had an intact copy of the estrogen receptor "beta" gene. I expect that in the future a large number of additional genes that are "turned on" by estrogen will be brought into the story as important for successful maternal behavior.

Finally, as animal behaviorists know well, individual mammals can inherit different proclivities toward parental care. We clearly see this phenomenon, for example, in different genetic strains of mice, which demonstrate different qualities of maternal behavior. But might some of the supposedly inherited influences be passed on within a family by new mothers simply mimicking the behavior of their own mothers? The question could be settled by what are called "cross-fostering" experiments, in which baby mice of one strain would be raised by the mothers of another strain. Such experiments, which would distinguish the genetic effects from those of the mother's behavioral example, have not yet been carried out. However, even in the absence of this information, we already know of half a dozen genes that definitely affect maternal behavior, as has been described above. These genes generate an ample number of candidate molecules that could account for the suppression of the "self-other" distinction, which, according to my theory, is required for ethical behavior.

Fatherly love

A father's love toward his children, just as the love by a mother, can lead to the emergence of positive, ethical behaviors. As Michael Numan and Thomas Insel point out in their excellent *The Neurobiology of Parental Behavior,* even male fish and birds may do an outstanding job of protecting the nest where babies are beginning to grow. Moving on to mammals, the authors discuss why males would provide parental care at all, considering that their investment in an individual female's pregnancy has been minimal. Since the male is just a source of sperm, so to speak, why not leave the mother to her own devices and go off to mate with as many other females as possible? Numan and Insel note that high levels of paternal care among mammals are usually found among species that show monogamous social organization. Therefore, in Numan's words, "the male improves his reproductive success by caring for young that are his own."

Since among humans the father as well as the mother may care for the children, the same principles we have discussed so far may apply to the behavior of good, caring fathers. Some men even undergo hormonal changes in response to their mates' pregnancy and birth: their level of testosterone falls after the woman has given birth, and the level of prolactin rises. These trends tell us how close to "maternal" some fathers can be, reinforce our view of humans as a two-parent species, and support the idea that both mothers and fathers could have contributed to the evolution of the biological mechanisms behind the Golden Rule.

Which genes contribute to successful paternal behavior? One leading gene codes for a hormone called vasopressin, the one that I already discussed in connection with stress and that is very similar to oxytocin. It is produced by certain large neurons in the hypothalamus and sometimes also in the amygdala. In a fascinating series of studies, Insel, Larry Young, and their colleagues at Emory University have found that the gene for one type of a vasopressin receptor has different patterns of activity in the brains of two different species of voles (tiny meadow creatures resembling mice), one monogamous and the other distinctly not monogamous. In turn, Insel showed that vasopressin is important for the forming of partner preference in these voles, and that this hormone seems just as important for the initiation of paternal behavior in males as oxytocin is in females.

One might ask, how could vasopressin, which is also associated with aggressive behaviors (as we will see in greater detail in Chapter 8), support caring, paternal behavior? The answer has two parts, one behavioral and the other molecular. First, part of the father's important role in raising the young is defending the nest. After all, he is likely to be larger and more ferocious than his mate, the mother, who is burdened with the job of nursing and feeding the young and might well feel depleted. A hormone like vasopressin, which fosters aggression, might be just what is needed. Second, at the molecular level, vasopressin binds not only to its own but also to oxytocin's receptors, almost as well as does oxytocin itself. Thus, it could exert some of the same effects on behavior as oxytocin, including initiating fatherly behavior.

A new twist in the relationship between genes and behavior emerged from a report published by Elizabeth Hammock and Larry Young at Emory University in 2005. Having noticed that some fathers within their colony of voles showed much more paternal behavior than others, the scientists wanted to find out why. As noted above, Young had already tied certain aspects of social behaviors to the actions of a vasopressin receptor. Now he and Hammock found that father voles who spent more time caring for their young differed genetically from those who spent less time with their offspring: a DNA segment called a promoter, which controls the activity of the gene coding for the above-mentioned vasopressin receptor, was slightly longer in the diligent fathers. Hammock and Young also found the same behavioral differences in the next generation of voles, depending on whether the sons had inherited the longer or shorter version of the gene promoter from their fathers. These experiments support the notion that paternal behavior, apart from being genetically controlled, can be inherited.

Recent scientific work suggests that male steroid hormones such as testosterone may also play a role in male parental behavior. A leading researcher in this field, Gabriela González-Mariscal, from Cinvestav-Universidad Autónoma de Tlaxcala, Mexico, observed that, in some species of animals, a rise in testosterone levels in the father accompanying the progression of the mother's pregnancy is "consistent with the male's behavior of guarding its mate." Surprisingly, after the babies have been born, testosterone levels decline, a shift that may help to reduce the chance of the father's being aggressive toward the newborn pups.

When parenting goes bad

Sometimes we can learn just as much about a phenomenon when things go wrong as we do when everything works right. As we have seen, a multitude of anatomical, chemical, and genetic mechanisms must proceed smoothly for mothers and fathers to raise their babies successfully. To learn more about these mechanisms, it is worth looking at instances and studies of failed parenting. (A note: I'll be citing studies of bad parenting in female animals, because, in mammals, parental behaviors are generally stronger and more frequent in females, and therefore, the failures are all the more striking and have been studied most extensively.)

In many animal species, mothers under pressure may abandon their young, fail to protect them, or even kill them. Our newspapers regularly report on disastrous failures of parental care in the human society. Children drowned, babies left in garbage cans, and many other heartrending stories remind us that the biological imperatives toward parental behaviors are fallible enough to be sometimes undermined by economic and psychological exigencies. Child abuse and child neglect, according to the Numan and Insel book, might be carried out most frequently by women who have a high level of depression and anxiety. That is especially true if the situation is aggravated by difficult circumstances that lead to frustration instead of the emotional rewards associated with child care.

The nerve cell circuitry responsible for parenting extends to several brain areas. Certain neurons in a region of the forebrain known as the medial preoptic area are crucial for normal maternal behavior. Damaging these neurons reduces all aspects of maternal behavior, regardless of the experimental conditions. Laboratory animals that have undergone this neuronal damage do not retrieve their pups in a well-organized manner, build a nest for them, or nurse them; instead, they neglect the pups for long enough to expose them to attack or starvation. These neurons are hormone-sensitive, as is clear from the work of Numan, of Boston College, and Susan Fahrbach, in my lab at Rockefeller: both scientists were able to deliver very small amounts of diluted estradiol, a form of estrogen, to these cells and show that maternal behavior was stimulated. It is in fact the hormone sensitivity of these neurons that accounts for the effects of hormones on maternal behavior.

The preoptic neurons do not work alone. They are connected to a circuit that travels from the preoptic area to the brain stem. Cutting the axons in this circuit, just as in damaging the preoptic area's cell bodies, reduces the performance of maternal behavior. Other influences on maternal behavior have been identified as well; for instance, large lesions of the cerebral cortex interfere significantly with maternal responses. This loss of cerebral cortical function in experiments can also lead to a loss of the sensory functions, preventing the mother from recognizing her young. In such a case, she may actually become afraid of them. In another part of the forebrain called the septum, damage does not abolish maternal behavior but does disorganize it. For example, the mother may carefully pick up the pups but distribute them randomly rather than bringing them to a single, well-organized nest. (Discussing the biology of fear in Chapter 3, I said that damage to the septum of the male can cause a much more extreme effect, known as "septal rage," but milder cases of septal damage to the female can result merely in disorganizing maternal behavior.)

Nerve cells in both the preoptic area and the septum have axons that, as they travel toward the brain stem, encounter a midbrain region called the ventral tegmental area. This is a most intriguing finding, because electrical stimulation of the ventral tegmental area has been found to be extremely rewarding for the laboratory animals in which it has been tested. (Electrical stimulation mimics natural neuronal communication. Neurons normally dispatch their signals to other cells by means of electrical jolts generated by neurotransmitters. Electrical stimulation makes the neuron produce its particular signal without a neurotransmitter.) Might this connection provide the intrinsically rewarding feelings that successful parental behavior can confer on us? And conversely, decreased activity in the preoptic and tegmental areas—along with increased fear-related output from the amygdala, as will be elaborated below—can probably exaggerate the woman's susceptibility to falling into a pattern of child neglect or abuse.

The part of the forebrain we saw lend emotional color to fear has effects on maternal behavior that are exactly opposite to those of the preoptic area: In the amygdala, the nerve cells inhibit expressions of maternal care. Numan, an expert in this area, feels that the amygdala's opposition to maternal behavior has a lot to do with fear. He calls the amygdala part of a "general aversion system," whose importance for maternal behavior might be twofold: if the amygdala's activity were not damped down, the new mother would actually be afraid of the strange new pups and would be more easily frightened by environmental stimuli that could be considered threatening.

This aspect of the amygdala's function takes on added meaning for women, as Numan points out in *The Neurobiology of Paren-*

tal Behavior, in a discussion "Hormonal and Neural Basis of Human Maternal Behavior" (Chapter 9), which lists the characteristics of women likely to foster the most reliable, successful maternal behavior. Those characteristics are all, so to speak, "amygdala friendly": they include a genetic makeup that favors a confident, low-anxiety personality; a secure relation with one's own mother; a normal hormonal background; and a low-stress pregnancy and ongoing environment. Numan posits that fear and anxiety militate against effective maternal behaviors. His reasoning can explain why the fear-signaling neurons in the amygdala would interrupt maternal responses if they are activated and would foster maternal behaviors if their activity is reduced. Thus, all the neuroscience pertaining to our fear systems that I described earlier in this book becomes relevant again—this time to explain how maternal behavior can sometimes be impeded.

Since estrogens powerfully support maternal behaviors, we might expect female mice that are missing a key estrogen-receptor gene to be absolute disasters as mothers; and so it proves. When provided with newborn pups they are supposed to "adopt," these mice not only fail to retrieve and feed them, they actually attack and destroy them. Moreover, this genetic deficit is likely to make itself felt in these females' brains quite rapidly, because when estrogen receptors are functioning properly, we know that their effect is fast: they enter the cell nucleus and alter gene activity as soon as estrogens bind to them.

Not all the genes known to be involved in maternal behavior actually support it—a few exert the opposite effect. In rats, opioid peptides—natural compounds with morphine-like activity—have been found to prevent the onset of maternal behaviors. This may seem surprising at first, as opioid peptides produce a soothing and pain-killing effect that should foster mothering, but it is confirmed by additional evidence: if opioid receptors are blocked, the disruption in maternal behavior caused by morphine-like drugs is abolished.

Moreover, the morphine-like drugs work on the same preoptic neurons I highlighted above as being crucial for the initiation of maternal behavior. Barry Keverne, director of the Sub-Department of Animal Behaviour at the University of Cambridge, has theorized that opioid peptides in the brain might mimic the positive feelings associated with successful maternal behavior, thereby eliminating the motivation for the behavior itself. Different species may react to opioid peptides differently; in monkeys, for example, injecting an opioid receptor blocker actually reduces the mother's caregiving and protective behaviors—an effect opposite to the one described above.

Do the genetic and neuronal cellular controls over parental and other social behaviors act in a precise and targeted manner? Insel and his colleagues, who have addressed this question, suggest that the answer is yes, and that even seemingly minor genetic and neuronal differences lead to major differences in behavior. Some of their most spectacular molecular findings have to do with DNA segments called promoters, which control the timing and pattern of activity of the genes coding for oxytocin and vasopressin receptors. Even in closely related animal species, the promoters for the oxytocin and vasopressin receptor genes are different enough for the gene to be active in different parts of the brain—a difference that, in turn, accounts for a major variation in social behavior, leading to (in the case of voles) monogamy in one species and polygamy in the other. Differences in the promoter for a particular vasopressin receptor gene also cause striking differences in brain activity patterns. Likewise, we know this genetic difference has real behavioral importance, because Young and Insel have already used their knowledge of the vasopressin promoter to develop a transgenic mouse—one carrying a different genetic sequence, transplanted from a different strain of mouse—and found that it exhibited more friendly, affiliative behaviors.

Summing up, I'd like to state once again my belief that many of these elaborate hormonal, neural, and genetic brain mechanisms, which evolved to facilitate reproduction, have subsequently become available to support a much wider variety of friendly, supportive behaviors that have nothing to do with sex or parenting. Once Nature comes up with ingenious ways of doing things, it does not throw them away: the mechanisms required for male-female courtship and for parental care are at the service of more complex social relations, of the sort needed to maintain the Golden Rule.

7

SOCIABILITY

Our lives link us with scores of other people, creating a myriad of contacts we tend to take for granted: the friendly runner who crosses our path every day at the same time, the neighbor with whom we discuss that new restaurant down the block, and the buddy who befriended us in the first week of college and who probably knows us better than we know ourselves. Yet our brains invest substantial resources in forming and maintaining these relationships—a striking reminder of how essential to us, how important an aspect of being human, it is to stay attuned to other humans.

How does the brain cope with the need to support the vast numbers of social interactions that hold our society together? As we have already seen, all forms of social bonding appear to stem from the same wellspring of the most basic human relationships. "I love him like a brother!" "She's been like a daughter to me all of these years." We all have heard these expressions, which speak volumes about the biological reference point for our varied social connections. If, as I propose, loving, supportive relations typical of stable sex partners and families *blend* into our feelings for friends and acquaintances in general, then the mechanisms of sexual and parental behaviors covered in the previous chapter should be able to tell us a great deal about mechanisms for the friendly, ethical behaviors that conform to the Golden Rule. That's indeed what happens, as we will see in this chapter.

Golden Rule hormones

Brain chemistry is coming close to explaining some of the biological mechanisms for sex, parental behavior, friendly behaviors in general, and, yes, love and trust. As the pieces of the puzzle fall into place, it gradually becomes clear that the intricate network of each person's unique emotions and social relations is critically dependent on a number of hormones that have a powerful impact on sociability. Since I believe that they also play an important role in prompting ethical actions, let us examine the research linking these hormones to behavior.

The major hormone of sociability—and a prime candidate for a chemical turning on the brain's "ethical switch" for friendly behavior—is oxytocin, a hormone produced in the hypothalamus at the base of the brain. It is a small molecule, consisting of only nine building blocks called amino acids, but its impact on our lives is enormous. As I mentioned in the previous chapter, oxytocin is essential for successful maternal behavior, but its role extends to a host of other social interactions in both women and men. According to Barry Keverne of the University of Cambridge, although friendly behaviors dependent on oxytocin may originally have evolved to support maternal care, eventually many other sociable behaviors, especially by females, have come to depend on this hormone.

Oxytocin and its receptor affect sexual, maternal, and generally friendly behavior in different ways, depending on the part of the central nervous system in which they carry out their activity. Oxytocin is produced by two groups of neurons in the hypothalamus, the paraventricular nucleus and the supraoptic nucleus, and distributed via nerve fibers to several major targets. First, by means of nerve fibers reaching to the posterior pituitary gland, oxytocin circulates in the blood and causes smooth muscle to contract, for example, in the uterus during labor. Second, oxytocin is transmitted to the lower brain, where it powerfully controls the autonomic nervous system (which is responsible for the automatic, or unconscious, regulation of internal body functions, such as blood pressure). At the same time, oxytocin receptors on spinal cord neurons control the sympathetic nervous system (which governs the smooth muscles, such as those that cover our blood vessels, viscera, and heart), and the administration of oxytocin increases the transmission of electrical impulses by these neurons. But most important for this book are the powerful effects of oxytocin on the neural networks of both maternal behavior and social behavior in general.

The essence of motherhood

Cort Pedersen, at the University of North Carolina, did the pioneering work to establish the importance of oxytocin for maternal behavior in female rats. Of all the brain chemicals he tried, oxytocin was by far the most effective at stimulating high levels of maternal behavior, and at doing so with a promptness that would enable the mother to protect and nurture her pups effectively to ensure their survival in the wild. Oxytocin-treated female rats engaged in a full range of maternal behaviors about four times as frequently as the control animals, ones that had been handled in an identical manner, but treated only with saline water.

Pedersen's experiments showed that oxytocin was sufficient for maternal behavior, but was it necessary? Across several laboratories, experimental results have shown that oxytocin is crucial for initiating maternal behavior, but not necessarily for maintaining it after it has been established. Susan Fahrbach, in my lab at Rockefeller University, used an oxytocin antagonist—a compound chemically similar to oxytocin that would tie up the receptor and block the normal action of the hormone—and found that it lowered the level of maternal behavior. Likewise, we found that maternal behavior was reduced if we introduced oxytocin-blocking molecules into the cerebrospinal fluid bathing the brain so that oxytocin would be scarfed up and unavailable for behavioral actions.

More evidence for oxytocin's crucial role comes from the laboratory of Thomas Insel, now director of the National Institute of Mental Health. Knowing that the oxytocin important for maternal behavior is produced by neurons in the paraventricular nucleus of the hypothalamus, he used electrolysis to damage those neurons in rats; the animals so treated then showed significantly less maternal behavior than their untreated littermates. Pedersen's subsequent work showed that especially important are oxytocin actions in the preoptic area and the ventral tegmental area—just the two regions of the forebrain I described in the previous chapter as being vital to the brain circuitry of maternal behavior.

How does oxytocin encourage the onset of maternal behavior? In *The Neurobiology of Parental Behavior,* Michael Numan and Insel propose that in some regions of the brain, oxytocin helps to decrease fear, thus reversing the female's avoidance of newly born pups and letting her retrieve them gently and return them to the nest. In other brain regions, oxytocin increases the mother's motivation to perform maternal tasks. And of course oxytocin stimulates not only the contractions of delivery in the uterus but also the ejection of milk from the mammary glands. All these disparate actions, taken together, illustrate the unique role of brain chemicals called the neuropeptides, such as oxytocin: their job includes orchestrating behavioral responses that chime with the physiology of the rest of the body, outside of the brain.

It's quite possible that one way by which oxytocin brings about maternal behavior is by reducing anxiety. Margaret McCarthy of the University of Maryland showed that only in the presence of estrogens, which rise sharply in pregnancy and after delivery, can oxytocin reduce anxiety in experimental animals. I assume the importance of the estrogens in McCarthy's experiments has to do with increasing transcription of the oxytocin receptor gene so that the receptor can be stimulated by oxytocin. Furthermore, oxytocin works by binding to a receptor in the membrane of responsive neurons, and interestingly, Insel showed that at the time that female rats give birth, oxytocin's binding to its receptor increases in some parts of the hypothalamus but not elsewhere in the brain—which suggests that around delivery, oxytocin may be selectively affecting neuronal pathways that reduce anxiety and bring on maternal behavior.

What is relevant for our discussion is that an understanding of oxytocin and other brain chemicals has implications beyond motherhood. North Carolina's Pedersen and others have proposed that the same neural circuitry and brain chemistry that evolved to provide the maternal care necessary for helpless mammalian infants also encourages adults to bond with one another. Numerous studies back up this claim.

Trusting oxytocin

Though we started the discussion of oxytocin with its vital role in maternal care, one should not get the impression that this is a purely female hormone. Some authors would say that oxytocin is more important in females and that the closely related hormone vasopressin, which we discussed in the preceding chapter in connection with paternal behavior, is more important for social functions in males. However, oxytocin receptors are widely distributed in both sexes, in the brain as well as in the spinal cord—as, indeed, are vasopressin receptors. Both oxytocin and vasopressin perform physiological roles associated with body fluids, oxytocin for lactation and vasopressin for responses to bleeding. But their social be-havioral roles, while overlapping, are not identical. Vasopressin is more connected with social aggression than oxytocin and may be more important in males.

Hormonal studies of sociable behavior have focused predomi-nantly on oxytocin, not vasopressin, but it's worth noting that the two hormones are close cousins and have probably derived from the same ancestral hormone, which scientists have named vasoto-cin. The work of Dietmar Richter and Evita Mohr, at the University of Hamburg, Germany, has suggested an interesting way that vasoto-cin could have given rise to both oxytocin and vasopressin: an "ac-cident" that occurred during the duplication of genes. As the vaso-tocin gene duplicated, mutations in two different parts of the gene may have led to the two mammalian brain hormones we know now.

Oxytocin encourages a wide variety of friendly, sociable behav-iors. Carol Sue Carter, at the University of Illinois of Chicago, and her colleagues have documented in laboratory animals the many positive social effects of oxytocin: more social contacts, higher rates of pair bonding, and decreased interference from stress or pain.

In humans, oxytocin increases feelings of trust, as Ernst Fehr and his colleagues at the University of Zurich, Switzerland, showed in a study in which volunteers played a game for money. In this game, each of two subjects plays the role of either the "investor" or the "trustee." If "Investor A," let us say, decides to trust (unidenti-fied) "Trustee B" and give money to him, the total amount available to the two subjects goes up. Trustee B can either share the mone-tary increase that, after all, is a result of Investor A's willingness to trust him—in which case both players end up with a greater pay-off—or he can keep the proceeds. The bottom line: the adminis-tration of oxytocin to the investor by means of a nasal spray signifi-cantly increased the investors' feelings of trust, more than doubling the percentage of those who transferred the maximum amount to their trustee.

The relation between oxytocin and trust holds true in reverse as well. If we perceive trustworthy behavior directed toward us, our oxytocin levels go up. Paul Zak and his colleagues at the Center for Neuroeconomic Studies, at Claremont Graduate University, experimented with a game similar to that used by Fehr. In this version, the index of trust shown by player A toward player B is the amount of money the former player is willing to send to the latter. Likewise, B's trustworthiness is shown by the amount of money he transfers back to A. The Claremont researchers found that when trustees received intentional signals of trust from investors along with the money, the trustees' blood levels of oxytocin were significantly higher than those of a control group of trustees who had not received signals of trust. Furthermore, trustworthy behavior, as measured by the amount of money that was returned to investors, was higher in trustees who had higher blood levels of oxytocin. Thus, higher oxytocin levels were linked both to being more willing to trust and to inspiring more trust in others.

These experiments all predicted happy results, but the obverse also applies: a lack of emotional care can produce an unhappy outcome. Seth Pollack and his team at the University of Wisconsin have published a study of the levels of oxytocin and vasopressin in children who had grown up in two different kinds of emotional environment. Children raised with normal levels of care displayed a rise in oxytocin levels when interacting with their mothers. However, children who had been neglected emotionally—they had spent the first few years of their life in orphanages—did not show this rise; moreover, they showed low levels of vasopressin. In Pollack's view, these abnormalities in oxytocin and vasopressin could account for social difficulties experienced by the once-neglected children.

As we have already seen with maternal behavior, in other social interactions oxytocin seems to produce its effects by decreasing fear and additional sensations that undermine social contacts, such as stress and anxiety. Thus, research conducted by Andreas Meyer-

Lindenberg, at the National Institute of Mental Health, suggests that oxytocin could encourage sociability and trust by decreasing social fear. In a brain-imaging study inspired by Fehr's work, he had volunteers take either oxytocin or a placebo by nasal spray. During the imaging procedure, the volunteers looked at angry faces and threatening scenes—sights that were frightening enough to activate the amygdala, the brain structure that we have seen prominently involved in fear and will revisit later in this chapter for its involvement in social recognition and friendly behavior. The main result of this study, however, was that oxytocin markedly dampened the activation of the amygdala as compared with the placebo treatment.

Animal research supports oxytocin's effects on the amygdala. In a study of male rats led by James Koenig at the University of Maryland School of Medicine, oxytocin injected into the amygdala reversed a chemically induced shortage of social interactions. These interactions had been reduced by more than 70 percent by drugs that block the normal action of the neurotransmitter glutamate. Oxytocin injections completely reversed this effect, restoring the social activity of the rats back to its regular levels.

Stress can take the fun out social bonding like no other state of body or mind—imagine trying to tell a joke to a policeman who's pulled you aside for a traffic violation—so it is little wonder that several ways in which oxytocin facilitates social behaviors may have to do with reducing stress. Scientists in the laboratory of Colin Ingram at the University of Bristol Medical School in Bristol, England, found that the action of oxytocin in the brain reduces the release of stress hormones from the adrenal glands. Inspired by several studies in which the administration of oxytocin directly into the brain reduced anxiety, they dosed rats with estrogen and progesterone levels that would mimic a late stage of pregnancy. When the scientists lowered the progesterone level to bring on maternal behavior, they saw an increase of oxytocin binding in the animals' brains and a decrease in anxiety. To confirm that the decrease was indeed due to oxytocin, they chemically blocked oxytocin activity and saw that the anxiety levels shot back up. Moreover, Inga Neumann, now at the University of Regensburg, Germany, found that blocking the action of oxytocin leads to an increase in hormonal responses to stress.

Finally, oxytocin could help us to be sociable in tense circumstances simply by calming us down. Most of my thinking has concerned the direct behavioral effects of oxytocin and vasopressin on the brain, but I also must take into account that, through their action on neurons in the central amygdala, both oxytocin and vasopressin influence our autonomic nervous system, affecting our heart, our breathing, and, literally, our gut feelings. Those indirect actions clearly could in turn influence our social proclivities. Margaret McCarthy and her research partners at the University of Maryland have shown that in the presence of adequate estrogen levels, oxytocin can act as an anxiolytic—it relieves anxiety, which is related to autonomic functions.

Where estrogens come in

Oxytocin's role in friendly behaviors brings us back to estrogens, as the effects of the two types of hormones are closely intertwined. Though estrogen—which, as we saw earlier, is involved in sexual and maternal behavior—may not qualify for the role of a "Golden Rule hormone," its close association with oxytocin makes it relevant for understanding the biological origins of ethical behavior.

The neuronal systems for oxytocin are bound tightly to estrogen's actions in the brain. Animals that have been surgically treated to produce very low amounts of estrogen also tend to have very little oxytocin function in the brain. When researchers reverse their estrogen deficit by giving them the main version of estrogen, estradiol, the level of their oxytocin activity in hypothalamic neurons crucial for brain arousal and behavior goes up because estradiol turns on the oxytocin receptor gene. Estrogen produces this effect by binding to the nuclear protein called estrogen receptor–beta;

we know this because Masayoshi Nomura, in my lab at Rockefeller University, discovered that estrogen's effect is abolished in animals whose estrogen receptor–beta gene has been deleted. In fact, Paul Shughrue and Istvan Merchenthaler, then working at the Women's Health Research Institute at Wyeth Research, obtained microscopic images of neurons with activated genes coding for both estrogen receptor–beta and oxytocin. With the giving of estrogen, oxytocin activity increases in the hypothalamus, not only in neurons that have fibers projecting to the brain stem and spinal cord to affect central nervous system arousal, but also in neurons that release oxytocin into the bloodstream for action throughout the body.

For reasons we have not yet understood, these actions of estrogen are opposed by thyroid hormones. What we do know is that certain hormonal systems are exceptionally slow, as some hormones take a long time to be effective, and they cannot be stimulated rapidly simply to encourage friendly supportive behavior during a single social encounter. Rather, they set up a physiological background that allows an animal or a person to respond positively to a given stimulus at a later date, given adequate social input. For example, changes in hormone levels during different seasons of the year affect the animals' willingness to engage in social or aggressive behaviors and to mate.

Estrogen's effects on hypothalamic neurons that produce oxytocin is most important, as estrogen also increases the activity of the gene coding for the oxytocin receptor. Vanya Quinones-Jenab in my lab used a specialized technique to demonstrate that estradiol "given back" to animals whose ovaries had been removed led to a huge increase in the production of the oxytocin receptor by hypothalamic neurons. Both oxytocin and its receptor show a rise in concentration after estrogen treatment, so the two increases may actually multiply each other and lead to a huge swell of oxytocin signaling. Janet Amico and her colleagues, at the University of Pittsburgh School of Medicine, carried out an impressive set of experiments showing that estrogen may trigger a rise in oxytocin as females come near giving birth. She mimicked the changes in sex steroid levels that occur toward the end of pregnancy: after introducing high levels of estrogens and progesterone into laboratory animals, she withdrew the progesterone and observed that the activity of the oxytocin gene went up significantly.

Researchers have also found evidence of a close link between estrogen and oxytocin by observing the electrical signaling of particular neurons. Yasuo Sakuma, now at the Nippon University School of Medicine in Tokyo, Japan, measured electrical discharges from oxytocin-producing neurons in the paraventricular nucleus of the hypothalamus of female rats who had been surgically operated on to remove their own sources of estrogens. When the scientists administered a particular form of estrogen experimentally to these animals, the electrical excitability of those oxytocin-producing neurons was increased. Within the paraventricular nucleus of the hypothalamus, this hormonal effect was narrowly targeted to oxytocin alone: the estrogen affected only the oxytocin-producing but not the vasopressin-producing neurons.

Blurring for friendship

How have we gotten from simple sexual bonding and maternal care to the tremendous variety of social relations that have a positive, cooperative nature and help to enforce ethical behaviors? Important as they may be, hormones obviously do not tell the whole story. Barry Keverne of the University of Cambridge and Carol Sue Carter of the University of Illinois at Chicago, who separately have extensively studied the social effects of oxytocin, would argue that the most primitive social behaviors connected with reproduction depended on hormonal controls, but that in higher species with more elaborately developed brains, social relationships can vary and change over time in subtle ways that cannot be directed simply by hormones. Moreover, with our human capacities for appreciating an enormous variety of environmental realities, the neural dynamics that control human social behaviors have been emancipated from a slavish devotion to reproduction and sex that dominate behavior in lower species.

In a wide range of human circumstances, including those governed by moral and religious teachings but not limited to them, social experience is free to take over the development of our social attitudes. Large parts of our big brains can become engaged in learning the rules by which we make each other's lives easier in society. We begin to learn early in life: the Swiss psychologist Jean Piaget famously observed that children learn from each other how to play fairly. To take Piaget's observation further, younger children learn from older ones, and the rules help both because they ease social relations. Many sociologists would hold that humans simply want to get along with each other and will do so if circumstances allow.

What actually takes place when people (or animals, for that matter) form a new social bond or strike a new friendship? First, we must learn the other's identity so that we can remember and recognize it. Second, assuming that the other person has not harmed us—no danger there—we recognize him or her as a friend. Then we can blur that person's features, his or her identity, with our own. We can "love that person as ourselves."

The path toward recognition

Since the identification of another person plays a major role in social bonding, I must ask: how do we recognize other individuals for who they are? We are beginning to piece together the molecular basis of social recognition through brain research on laboratory animals.

A discussion of the evolutionary history that has led to social bonding among members of monogamous species has been launched by Barry Keverne. In his view, many nonhuman primates would not survive if they depended on the bond between mother and infant alone. Although social bonding on the whole owes its signals and signal pathways to those genetic and neural steps that permit maternal behavior, by themselves, these steps are not enough. Cherished relationships must also extend beyond the immediate family, to permit living in larger social groups. Turning to specific brain mechanisms, Keverne points out that as our ancestors' brains grew larger in the course of several million years and our distance senses—that is, sight and hearing—gained in range and power, our sense of smell became less important for social recognition and bonding.

Animals, on the other hand, obtain much of the social information they need through their noses. Because virtually all pheromones, the substances animals secrete to affect the behavior of their peers, and other odor molecules send their signals through the forebrain pathways that lead to the amygdala, this collection of neurons once again comes into play. We have seen how key the amygdala is in our fear circuitry—and I also listed it as a candidate for harboring the identity-blurring mechanism behind ethical behaviors—so it seems significant to find real evidence of its role in social recognition as well.

To investigate the mechanisms involved in such recognition, Elena Choleris, I, and others in my lab at Rockefeller University analyzed the ability of female mice to recognize other females. Choleris used very precise laboratory techniques that told us not only how the test females got used to females they recognized—and knew that those intruder females did not represent any threat—but also how the test females would renew their investigations when a new, strange intruder female was introduced among them. We studied genetically normal mice and then compared their social recognition performance to that of littermates in which either the oxytocin gene or one of two sex hormone receptor genes had been deleted, or, in technical language, "knocked out." We had chosen to manipulate these particular genes because in lower mammals, many aspects of social behavior appear to be affected by hormone-dependent systems connected with reproduction. Indeed, we found that "knocking out" any one of these three genes significantly reduced all aspects of social recognition. When we find several genes work together to produce an important behavior, we call this a "gene micronet."

Choleris used these results to propose a convincing model explaining that social behaviors are linked to reproduction, just as Keverne had anticipated and as I theorized above. According to her reasoning, the female mouse's ovaries secrete estrogens as the animal is getting ready to ovulate. Circulating in the blood, the estrogens, retained by neurons in the hypothalamus, turn on the gene that codes for oxytocin in those neurons. The elevated levels of oxytocin are transported to the amygdala. At the same time, the estrogens, having circulated in the blood, also are retained by neurons in the amygdala, where they turn on the gene for the oxytocin receptor. Thus, the elevated levels of oxytocin receptor are ready to receive and react to the oxytocin transported from the hypothalamus. As a result, oxytocin's activity soars, greatly increasing the animal's sociability. Choleris emphasizes that the *concurrent* activity of these various genes in their *different* locations in the forebrain would be crucial in order for social recognition to work correctly.

It is significant that these molecular events take place in the amygdala, for two reasons. First, it is to the amygdala that odor and pheromonal signals are transmitted, providing the basis for social recognition. Second, it is precisely in the amygdala that oxytocin, working through the oxytocin receptor, fosters increased social recognition. Note how the amygdala keeps emerging as a brain structure that is key to upholding the Golden Rule!

Jennifer Ferguson, working with Thomas Insel, found that injections of oxytocin to the amygdala improved social recognition. Choleris and I tested the converse proposition, using a special molecular trick called antisense DNA, which prevents the gene from making the protein it codes for. In this way we were able to block

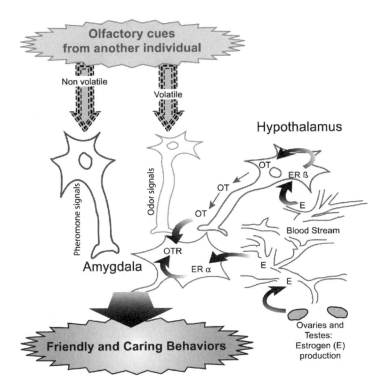

Figure 8. A four-gene micronet. Genes for estrogen receptors (ERα and ERβ) and genes for oxytocin (OT) and its receptor (OTR) form a "micronet" controlling social recognition and, thus, friendly and caring behaviors.

the activity of the oxytocin receptor gene in the amygdala, and as a consequence we saw a decrease in social recognition. All these findings provide us with a rather comprehensive understanding—from molecular chemical details to brain anatomy to animal behavior—of how oxytocin and oxytocin receptor function in the amygdala to foster social recognition in mice.

I argued earlier that these primitive molecular and anatomical relations have been retained in the human brain and operate there in much the same way. Of course our social relationships depend on myriad cultural habits and customs overruling the primitive, sexy drives, and our recently evolved human cerebral cortex overrules the human amygdala. Still, the understanding gained from animal studies sheds light on the roots of human social behavior.

Is there a gene for social recognition? In our current understanding of the genetic basis of behavior, there is certainly no single gene accounting for this trait. Choleris's model of social recognition, even in laboratory mice, does not invoke a simpleminded claim of "one gene/one behavior." That old idea clings even today because decades ago, in the time of classic genetics, George Beadle and Edward Tatum won the Nobel Prize for the "one gene/one enzyme" concept they had developed from their study of *Neurospora*. But modern neuroscientists have moved beyond that. In my lab, for example, in explaining mechanisms for behaviors that are different between males and females, we have shown that *patterns* of genes govern *patterns* of behaviors.

Nor does Choleris try to wrap up the explanation of social recognition and altruism in a single gene, as these two actions might have different genetic roots. A mathematical approach reported in *Nature* in 2006 by Vincent Jansen and Minus van Baalen, at the University of London, supports this notion. The scientists used a computer model to simulate what would happen in a typical game of Prisoner's Dilemma in which, as described earlier in this book, two players must decide to implicate or keep quiet about the other's involvement in a fictional crime. Assuming that social recognition and altruism are always inherited together, Jansen and van Baalen mathematically tested the implications of the simplest genetic possibility: that social recognition and altruism both depend exclusively on the same gene. Their calculations reveal that such an arrangement would make most social relationships highly unstable. Cooperation would bounce from absent to present to absent again, in a fashion that could not support a normal society, animal or human. If, instead, social recognition and altruism were caused by what Jansen and van Baalen termed "loosely coupled separate genes," whose activity could influence traits, but not determine them absolutely, then a variety of recognizable features across the population would greatly foster altruistic behavior. This concept of the loose coupling of genes is important, because it provides a genetic mechanism by which people can recognize each other as altruistic and behave appropriately even if they are not related to each other.

The circuitry of friendship

Scientists have designed several ingenious experiments to study the workings of friendship in the human brain. The research team led by Tania Singer, at University College London, is using functional magnetic resonance imaging to reveal some of the nerve cell groups whose activity is correlated with empathy of one human for another. They asked volunteers to play a game with employees of the lab, secretly instructing the employees to play either fairly or unfairly. Afterward, the scientists measured brain activity in the same volunteers under quite different circumstances: looking on as their former game opponents were subjected to various degrees of pain. In both male and female volunteers, the brain areas that signal pain became active, giving neural evidence of their empathy with the others' pain. Strikingly, however, that empathy did not appear to extend to all the players who were hurting. When unfair-playing employees were seen experiencing pain, the male volunteers—but not the females—showed significantly less empathetic brain activity than when they saw fair-players receiving pain. Thus, females showed the brain responses of empathy regardless of their moral judgment of the employees' social behavior, whereas the men's brain responses were conditional upon how fairly the employee had played. Here is rich material for a discussion that, unfortunately, falls outside the scope of this book, but for the purpose of our argument, this study provides interesting insights into ways in which the notions of empathy and fair play are represented in the human brain.

Singer also asked the obverse question: how does the human brain respond to the positive aspects of reciprocal cooperation? In this experiment, volunteers played the above-mentioned Prisoner's Dilemma game and were told to keep in mind the faces of people who cooperated and those who did not. Then the volunteers were shown images of all the faces and asked to rate how well they liked them. The faces of likable cooperators gave rise to increased activity in the left amygdala (though, surprisingly, not the right—a finding we do not yet understand) as well as to the bilateral activation of other forebrain areas related to emotion and reward. Once again, the amygdala revealed itself as important in issues of social bonding.

In evolutionary terms, the brain circuitry supporting the relationships examined in Singer's studies could very well have developed from the nerve circuits of parenting, which melded into patterns of neuronal activity that produce a wide variety of social behaviors. The well-known neuroanatomist Sarah Winans Newman has pointed out that many different social behaviors seem to use the same neural circuits in the forebrain. This kind of overlap could make the melding of behavior even more likely.

One of the neuroanatomists who has worked longest and most systematically on this type of question is Larry Swanson, at the University of Southern California. He postulates that organized layers of signal-transmitting fibers running through the hypothalamus govern entire classes of behaviors. Some of these systems, for example, could be devoted to ingestive behaviors, like eating and drinking. Others might be devoted to reproductive and other social behaviors, while still others might have to do with aggression. One important implication of Swanson's thinking is that forebrain and hypothalamic controls over elemental social behaviors would not necessarily govern one particular response as opposed to another. Instead, sociability itself might be the subject of one of the neural systems in question. Such ideas reinforce my opinion that mechanisms supporting maternal and paternal behaviors blend into the controls supporting positive, friendly, social behaviors as a whole.

When we love, show compassion, or feel sympathy, we "identify with" our companion. And when we decide that we are going to be friendly to others, we have thus recognized, we are using ancient forebrain circuitry that we share with lower mammals.

From family to inclusive fitness

Networks of social behavior have evolved among all vertebrates, encompassing a range of animals from reptiles to amphibians. According to James Goodson and his colleagues at the University of California at San Diego, even fish display the hypothalamic-amygdala relations that have been investigated by Choleris and described earlier in this chapter. Goodson has also found that in fish, social behaviors (in the form of courtship noises) are linked to their requirements for reproduction. Birdsong provides us with another obvious example. In many species, males sing much more than females and use their songs to broadcast the extent of their territory and to attract females. Most important for the main arguments of this book, bonding behaviors that start with reproduction and parenting and extend to other social situations, also apply to higher mammals such as nonhuman primates, and thence to humans. In the words of Robert Axelrod, at the University of Michigan, once the genes for cooperative behavior have evolved, natural selection will favor certain social behaviors and will lead to "strategies that base cooperative behavior on cues from the environment."

Dwelling on human cooperation in larger groups of people, the UCLA criminologist and social scientist James Q. Wilson sets cooperative, friendly human behaviors in the evolutionary framework established by Charles Darwin. Natural selection shapes the social behaviors of individuals who will survive to maturity, mate, and have a lot of progeny. To this, evolutionary biologists Robert Trivers and William Hamilton add a key insight: in some cases, an individual can propagate his genetic material into the future better by helping many members of his extended family (who share some of his genes, though not all of them) than by focusing his efforts exclusively on his own few children. An individual's genetic "fitness" includes not only his own ability to produce and raise offspring but also the ability of his kin who have some genes in common with him—the more, the better.

Wilson not only employs this thinking to understand why we exhibit positive, group-oriented behaviors, but also recognizes that whenever people treat each other in a fair, sympathetic manner, they are exhibiting an essential understanding of the importance of reciprocity. In his words, "The norm of reciprocity is universal." If we do a favor, we expect one in return. If we receive a favor we cannot return, we are distressed. Wilson takes his argument from the mechanisms of parental care to the basic human desire for attachment—a friendly social environment fostered by ethical, fair, sympathetic behavior.

The scholarly ethicist Stephen Post, of Case Western Reserve University, extends this line of thinking to ponder the nature of love. In his view, love yanks us out of our self-centered cocoons so that we can consider another person's interest. He recognizes that altruistic love springs from evolutionary strategies for survival, including the formation of the parent-infant bond. Thus we see a continuum from detailed brain mechanisms for mating and parental behaviors, through normal human cooperation, to love.

When this system fails, we've got trouble on our hands. Consider autism, the spectrum of developmental disorders that are, in the language of the National Institute of Mental Health, characterized by "severe and pervasive impairments in thinking, feeling, language, and the ability to relate to others." I suggest that the molecular, chemical, and neuronal mechanisms described in this chapter as required for both maternal love and social friendship are damaged in autistic children and we must find a way to repair them. The speculation of Meyer-Lindenberg discussed above, that oxytocin's ability to dampen social fear may aid the development of sociability, offers another perspective on potential therapies for autism.

Several decades of scientific study have made it abundantly clear that social bonding mechanisms are necessary for mental and physical health. In a now-classic set of experiments conducted by Harry Harlow on rhesus monkeys at the University of Wisconsin, a withdrawal of normal maternal care caused the young monkeys to isolate themselves, to communicate less with other monkeys, and even to slow down their rate of growth. This last symptom may be explained by a study from Judy Cameron, at the University of Oregon, and her colleagues: monkeys whose early life experiences have made them behaviorally inhibited—"shy," in everyday conversation—due to early experiences show significantly less responsiveness to growth hormone than a control group of monkeys.

Several authors have reported evidence that when we are excluded from an emotionally important relationship, the social affront to us activates neural circuits that are normally involved in the control of our guts and the perception of physical and psychological pain. Conversely, a supportive social environment can reduce the emotional impact of pain, in terms of its intensity and duration.

Throughout the range of all possible social interactions, neuroscientists are beginning to piece together the influences on neural networks that regulate sex, social recognition, and sociability. These networks, in turn, factor into the crucial steps supporting ethical decisions that I described in chapter 5. Wonderfully, as I proposed in that chapter, we do not need to posit increased levels of performance in these neural mechanisms, but instead can be confident that once we recognize another as a friendly, nonthreatening presence, then it is *decreased* social recognition that leads someone to obey the Golden Rule. A person momentarily forgets the difference between himself and the other, and as a result he complies with a universal ethical principle.

Thus the neurobiology for normal harmonious social interactions in animals and humans provides a set of positive mechanisms for ethical responses in neutral, unthreatening situations, such as the daily exchange of nods with the neighborhood runner or a long-awaited beer with old friends. And when the friendly hormonal mechanisms are not quite up to the task, the set of fear mechanisms covered in earlier chapters of this book will kick in to help us deal ethically with threatening, fateful situations.

Backed up by these two sets of biological support systems, most people respond to most situations in an ethical way—except, that is, when they don't. It would take, however, more than a minor disruption of the mechanisms of ethical behavior for the Golden Rule to fail.

8

THE URGE TO HARM

We have a problem. With the theory put forth in this book, I have been seeking to explain behavior that follows an ethical principle. News flash: as we know all too well, people behave unethically all the time.

What goes wrong? How can the Golden Rule be trumped so readily? Surprisingly enough, committing a scandalous act does not always mean throwing out the Golden Rule. For example, one form of illegality, not warlike or violent, has a peculiar relation to the ethical principle I have been discussing. From New York to Moscow to Macao, we hear about corruption: people getting favors from officials in return for illegal compensation. Why should this happen? The answer might appear shocking. Corruption, which seems to be just as universal as the observance of the Golden Rule, is nothing less but its extension—outside of the law. "You do something for me; I do something for you." With the corrupt duo, it is a form of reciprocal altruism, albeit illegal.

In no shortage of behaviors, however, the Golden Rule is obviously overturned. Take its starkest reversal: aggression that escalates into violence. Our newspapers are filled with reports of human beings harming one another, and the history of the human race is studded with wars between nations and attempts to wipe out an entire people. The Holocaust, which occurred in the mid-twentieth century, is often cited as the epitome of human atrocity, but acts of genocide are known to have occurred before and since; just in the past few decades, there have been at least five attempted genocides, in Africa, Europe, and Asia.

Is an urge to harm embedded in our brains? Is ours a vicious species? My theory postulates that the Golden Rule mechanism is fairly robust and supported by two profound systems, the fear system and that for loving behaviors. If aggression and violence can override this mechanism, they surely must have deep roots. What makes them so powerful?

The many different contexts in which humans behave terribly toward each other are subject to many different types of explanations—psychological, sociologic, historical, economic, and others. All rise from deep research and careful inference. Neuroscientific approaches complement scholarly understanding by supplying the deepest layer of explanation. Consider layers of explanation to be like an onion: neuroscience may be at the core, but it takes all the layers to make an onion.

Some neuroscientists and others boil the onion down to inborn instincts toward violence—a "killer instinct"—to explain calculated killing. While that is hardly a complete explanation, I do have confidence that causes of violence have neurobiological roots. Throughout evolution, both animals and humans have relied on aggression in a variety of situations, such as hunting for food or protecting themselves from being devoured by predators. Thus, aggression might confer an evolutionary edge under certain circumstances: the British biologist Richard Dawkins considers the simple defense of territory and retaliation against another's aggression as basic strategies that must have persisted due to the survival advantages they provide.

Yet whatever the reasons for aggression, a person committing a violent act fails to see himself in another, that is, the blurring of identity put forth by my theory fails to take place and all the mechanisms of ethical behavior fall apart or are at least temporarily subdued. Exploring the origins of aggression in all its forms, from self-defense to senseless self-destruction, prepares the ground for explaining what happens when violence triumphs over ethics, and when it does not.

War and peace

Before turning to the neurobiological causes of aggression and violence, lets us take a brief tour of additional perspectives provided by experts from other disciplines. In seeking to explain the recurrence of wars, scholars and thinkers from various fields place violent behaviors in the general context of the human experience. For example, the Prussian officer Carl von Clausewitz declared war "a continuation of policy by other means"—in other words, a perfectly rational undertaking.

In our own time, the social critic Barbara Ehrenreich, despite her Ph.D. training in biology, disputes the notion that a biological "killer instinct" has much to do with the causations of war. Among other factors, she writes, the aggressors in a war may hope to capture resources and thereby "advance their collective interests and improve their lives." For Ehrenreich, war is "too complex and collective to be accounted for by a single warlike instinct lurking within the individual psyche." Instead, she is interested in the ritualistic character of war and wants to identify "what it is about our species, about human nature" that leads us to kill other humans in a systematic, large-scale manner. Her answer lies in ancient blood rites, the earliest forms of organized violence for which we have evidence. In Ehrenreich's words, religious rituals of blood sacrifice, such as the sacrificial lamb, or even certain forms of self-mutilation, "both celebrate and terrifyingly reenact the human transition from prey to predator." These rituals laid the social groundwork for war, and the war methods then evolved over the centuries.

The cultural anthropologist Raymond Kelly, from the University of Michigan, takes a different approach. He seeks to explain the transition from warless societies to the practice of organized war more than 14,000 years ago. According to Kelly, maintaining a warless state of affairs depended on low population density, adequate measures of deterrence, and the appreciation of positive relations among neighbors. Then, he says, changes in social organization "facilitated the mobilization of all adult male group members" for the purpose of carrying out preplanned, warlike acts that enlarged territory and gained the victors more resources. His key concept is the notion of "social substitution." If male A kills someone in male B's tribe, is male B culturally required to kill someone in male A's tribe even if that intended A victim had nothing to do with the original violent act? If not, you have a warless society; but if he is so required, you have the kernel of "justified," organized violence that evolved into war. In a certain society in New Guinea it was believed that if male A had killed someone in male B's tribe, the spirit of that victim would hang around the B camp and make trouble for everyone until someone in tribe A had been killed. Then the same dynamic would apply in the camp of tribe A, and so on, and on. Endless violence.

The recurrence of violent warfare has been a major subject of historical studies. The historian Barbara Tuchman has a simple way of explaining how we stumble into wars. She cites the loss of rational thought in the service of what she memorably calls "political folly," that is, "pursuing disadvantage after the disadvantage has become obvious." For Tuchman, the main reason for the "persistence in error" is the lust for power and all the resources that come with it.

Donald Kagan, professor of history at Yale, examines the causes of war by looking for commonalities among four major conflicts separated by centuries. A student of what he terms "comparative history," he seeks common threads in the wars between Athens and Sparta and between Rome and Carthage more than two millennia ago, as well as in last century's World Wars I and II.

In analyzing the Peloponnesian War, between Athens and Sparta, Kagan rejects both the claim by the ancient Greek historian Thucydides that the war was "an inevitable result of the growth of the Athenian Empire" and the explanation by German scholars that it represented a formation and execution of foreign policy in order to deal with domestic problems. Instead, Kagan describes a "reasoned calculation" by the Athenians about the behavior of their fractious allies, the Corinthians; contrary to this calculation, however, the Corinthians provoked the Spartans, who then launched an attack. In other words, Kagan's view is that a miscalculation led the Athenians into a war they were trying to avoid.

The conditions for the Second Punic War, with Rome battling Carthage and Hannibal, were set when Rome won the First Punic War and caused resentment by refusing to ratify a reasonable treaty. Meanwhile, Hannibal was spoiling for a new war and had secretly established alliances with Celtic tribes in Europe. When Hannibal attacked a critical Roman base in Spain, it took Rome more than a year to respond, giving Hannibal time to invade Italy. Having first bargained too hard on a treaty and then failed to prepare effectively for the next war, says Kagan, Rome was forced to "pay the price of a long, bloody, costly, devastating and almost fatal war."

In the run-up to World War I, Germany was a strong military power dissatisfied with the status quo. Kagan notes that the British could have deterred the Germans if they had formed "formal and open military alliances with France and Russia." In contrast to the ancient Athenians, the British could have achieved deterrence but were unwilling to take the actions to do so.

The Second World War also derived, in part, from mistaken policy related to an earlier war. The European democracies were so affected by the horrors of the First World War that, according to Kagan, they held "a blind hope that a refusal to contemplate war . . . would somehow keep the peace." Had the European democracies prepared effectively for the possibility of war with Hitler, his generals might not have tried to follow his ambitious plans for conquest.

"No peace keeps itself," as Kagan says repeatedly. All the wars he cites arose from mistakes in social and governmental policy, not from cellular mechanisms in neurobiology.

Yet despite these valid and illuminating views, the ubiquity of violent behaviors, of personal and organized violence, does suggest that the capacity for this kind of aggression is biologically built-in at least to some extent. In fact, even theorists who would argue against an important role for a "killer instinct" recognize biological determinants of violence and war. In Barbara Ehrenreich's phrase, "War is in fact one of the most rigidly 'gendered' activities known to humankind." If "gendered," then it allows important roles for chromosomes, sex hormones, and their effects on brain function. One way to burrow into this question is to figure out what chemical, genetic, and structural mechanisms in the brain lead to aggressive behavior in animals and humans.

The chemistry of aggression

Strong support for the notion that human aggression has biological roots comes from well-documented evidence of organized attacks in other species. For example, communities of chimpanzees are known to occasionally attack one another: especially during long dry seasons when food is scarce, if a group of three or more male chimpanzees from one band sees a lone chimpanzee from another band, the three essentially form a war party and kill that potential invader. Richard Wrangham, at Harvard University, argues for similarities between organized acts of aggression among male chimpanzees—aimed at capturing resources and conferring social power—and the actions of humans who have generated "a system of intense, male-initiated territorial aggression" for the purposes of enlarging territory and killing enemies.

Neuroscience is already making progress in the search for mechanisms that can explain what happens when one living being harms another. We have come a long way since the days of classical ethology, the scientific and objective study of animal behavior in natural conditions. The famous book *On Aggression,* by Nobel laureate Konrad Lorenz, describes animal behaviors in a way that was dramatic and new to the world in 1963, but it does not go beyond description. Lorenz portrays normal forms of aggression that are biologically necessary—for example, self-protection—and are codified in a manner that limits the bodily damage likely to be suffered by a given individual. Now we can analyze the mechanisms underlying aggressive behaviors in animals and humans with the tools of molecular biology, physiology, genetics, and quantitative behavioral science. We are learning a great deal about aggression in the brain, about who may be aggressive, and a little bit about the nuances of aggression as well as its relationship to disorders and susceptibility to treatment. In particular, a number of hormones, neurotransmitters, and other body chemicals are gradually emerging as important players in the making of aggressive behavior.

Testosterone

A jibe at strife between competitive males—that women make sure men hear—goes something like, "They've got testosterone poisoning." Biologically speaking, that's a good guess: males are far more prone to violent aggression and testosterone's molecular fingerprints are everywhere—almost.

According to a 2006 report in the *New York Times,* of 1,662 slayings committed in the city over the preceding three-year period, 93 percent had been carried out by men between the ages of 18 and 40 years. Within American families, according to other studies, more than 90 percent of violent acts are initiated by men. In addition, the incidence of murder of males by unrelated males in American society, as well as in other societies, follows a lifetime curve that closely parallels that of the testosterone concentration in a man's bloodstream. It looks as though testosterone and other male hormones acting on the male brain predispose some men, in some situations, to commit violent acts. Therefore, the search for chemicals responsible for aggressive behavior first and foremost points to testosterone.

Back in 1990, Harrison Pope, at Harvard Medical School, was the first medical scientist to provide convincing evidence for the strong effects of male hormones, including testosterone, on human aggression. In a series of papers, he described a range of testosterone-related changes in mood and behavior, from feelings of hostility to such extreme acts of aggression as homicide.

The association of testosterone with aggressive behaviors spans cultures and geographical borders. In San Sebastian, in the Basque region, and in Seville, Spain, Aitziber Azurmendi and his colleagues found a highly significant correlation between testosterone and other male hormones in the blood and the tendency toward provocative behavior, even in five-year-old boys. An international team led by Irene van Bokhoven, of Utrecht University in the Netherlands, studied adolescent boys and saw similar correlations between testosterone and self-reported delinquent behavior, as well as proactive and reactive aggression. At age 16, according to the researchers' scientific paper in *Hormones and Behavior,* boys who had already developed a criminal record had increased blood testosterone levels.

Not only are testosterone levels in the blood correlated with aggressive behaviors, but also it is clearly established in a variety of experimental animals that castrating the male—that is, removing the major source of testosterone—drastically reduces aggression. In turn, injecting testosterone into the castrated male animal restores aggressive behavior.

In humans, consider a loss of testosterone versus an excess. Eunuchs have never been known for exhibiting physical aggression; on the other side of the coin are the numerous studies, including the ones mentioned above, linking increased testosterone and aggression.

One must not, however, jump to the conclusion that the relationship between testosterone and aggression is simple and straightforward; rather, complex mechanisms may be at play. Thus, as part of normal body metabolism, testosterone can be chemically converted into two different substances, called testosterone metabolites: a chemical known as dihydrotestosterone and—surprisingly enough— certain forms of the female hormone estrogen. (The latter transformation is part of normal body chemistry and does not at all mean that a man starts acting like a woman; explaining it in greater detail falls beyond the scope of this discussion, but I'm mentioning it here to provide an idea of the complexity involved.) As does testosterone itself, a combination of the two testosterone metabolites restores aggressive behavior in castrated male animals. Moreover, this combination mimics testosterone's effects on the brain's regulation of vasopressin, a hormone whose role in aggression is described below. Finally, male hormones flooding the testes might increase aggression by four different routes, summarized by Neal Simon, at Lehigh University: through the action of each of the three substances—testosterone and its two metabolites, dihydrotestosterone or a particular form of estrogen—on the brain, or through a mutually enhancing effect of all the three substances combined. These routes are not exclusive of one another, and they might have different relative importance in different males.

What are the mechanisms by which testosterone works to affect aggression? Some of these are indirect and do not involve the brain

at all. Testosterone and other steroids responsible for masculine characteristics make muscles grow; large muscles, in turn, would make a
guy more able to fight victoriously. They also would make him more
confident that he could win, and thus possibly encourage him to start
a fight. Put two of these guys together and you have trouble.

In the brain, testosterone, like estrogen, has both fast and slow
mechanisms of action, and these affect aggressive behavior in different ways. In the slow route, testosterone works by entering the cell
and binding to a large protein called the androgen receptor. Once
bound to testosterone, this protein enters the nucleus of the nerve
cell and attaches itself to specific portions of DNA that regulate the
activity of testosterone-sensitive genes. Slowly, over the course of
hours, the androgen receptor increases the activity of some genes
and decreases the activity of others. The resultant changes in brain
chemistry can gradually shift a person's mood and behavior into a
more aggressive mode. A second way by which testosterone affects
nerve cells is fast. John Nyby, at Lehigh University, and others have
reported evidence that testosterone can bind to receptor sites embedded in the membrane of nerve cells to influence the activity of
those nerve cells within seconds or minutes. This near-instant effect
could trigger a rapid change in a man's behavior resulting in an impulsive act of aggression.

Vasopressin

Another hormone thought to play a role in regulating aggressive behaviors reveals just how vital aggression might be for survival.
Vasopressin, which we have seen to support a father's caring behavior, also promotes aggression in males. This might seem a contradiction, but in fact the two behaviors are closely intertwined: as I
pointed out when discussing this phenomenon in Chapter 6, part
of the male's parental behavior is protecting the offspring against
attack to ensure their survival.

Several studies from the laboratory of Craig Ferris, at the University of Massachusetts, and Elliot Albers, at Georgia State University, have connected vasopressin with aggressive behaviors. Injecting vasopressin into certain parts of the hypothalamus lowers the psychological threshold for aggressive behaviors—in other words, it makes such behaviors more likely; injecting a chemical that blocks access to vasopressin receptors makes these behaviors less likely. Albers and Kim Huhman, also of Georgia State University, believe that when social manipulations increase aggression, one of the molecular mechanisms behind this effect is an increased binding of vasopressin to its receptors in the hypothalamus. And according to Ferris, in certain environmental contexts vasopressin enhances the arousal of the central nervous system, which in turn amplifies the aggressive behavior.

Most of these studies have been conducted in males, but it would be too simplistic to state that vasopressin promotes aggression in males whereas its close cousin oxytocin promotes sociability in women: as usual, the actual picture is much more complex. However, many experimental results do fit this statement, and vasopressin's tight connection to testosterone, the major male sex hormone, reinforces vasopressin's involvement in male aggression. Thus, part of the way in which testosterone increases aggression is through its stimulation of genes involved in carrying the chemical message transmitted by vasopressin. And conversely, the binding of vasopressin to its receptors in the parts of the hypothalamus where this hormone stimulates aggression depends on the presence of testosterone. Moreover, Geert de Vries, at the University of Massachusetts, an authority on the effects of male hormones on the brain, has found that males have more vasopressin-producing neurons in the forebrain and more signal-transmitting extensions from each such neuron than females.

Human studies, too, link vasopressin to male aggression. Richmond R. Thompson and Jeff Benson, working at Bowdoin College and Scott Orr of Harvard Medical School, respectively, looked at how vasopressin influences social communication in humans. The scientists measured the electrical activity of certain facial muscles of study participants to evaluate their response to strangers. When responding to pictures of the faces of unfamiliar men, male study subjects receiving vasopressin displayed more *un*friendly facial motor patterns than a control group. In contrast, women who received vasopressin displayed more friendly facial motor patterns when responding to the faces of unfamiliar women. The results shown by men fit in very well with the studies of animal brains and behavior, but the sex differences in human social behavioral responses were unexpected. If new research can extend these findings in additional studies, I would be tempted to point out they are consistent with the greater tendency of women, in general, to cooperate in a given situation.

Serotonin

Serotonin plays a broad range of roles in the body, from constricting blood vessels to regulating body temperature, sleep, and mood. It is also believed to be one of the major neurotransmitters regulating anger and aggression and a powerful inhibitor of aggressive behavior. However, serotonin is an important example of a neurotransmitter having different effects depending on the receptor subtypes to which it binds. The exact nature of the serotonin effect on aggression depends on the subtype of receptor through which the serotonin is acting. In fact, different receptor subtypes can exert completely opposite effects: deleting some subtypes can lead to an increase in aggression, whereas deleting others can decrease aggression.

The synthesis and breakdown of serotonin figure largely in the regulation of aggression, and the neurons that produce serotonin are important for governing aggressive behavior. These cells are in the midbrain, very close to the midline, in groups called the medial and dorsal raphe nuclei. *Raphe,* the Greek word for "fence," applies well here because these cell groups look like a fence separating the left midbrain from the right.

Stephen Manuck and his colleagues, writing in Randy J. Nelson's book *The Biology of Aggression,* describe their evidence that low levels of serotonin in the brains of rhesus monkeys caused the animals to have fewer social companions and spend less time "in passive affiliation" with one another or grooming their peers. The scientists observed a pattern that turned out to have wide significance: whereas monkeys with abundant serotonin in the brain displayed little aggression, perhaps only baring their teeth now and then, low levels of serotonin production in the brain often caused the animals to initiate intensely aggressive behaviors such as angry pursuit and real attempts to bite.

In humans, too, a shortage of serotonin is obviously linked to aggressive behavior. Manuck and his colleagues found that low levels of serotonin-related activity in the human brain clearly predispose the individual toward intense aggression, and Emil Coccaro, at the University of Chicago, has tied low levels of serotonin-related activity to long-lasting tendencies toward extreme forms of impulsive aggression in certain patients with personality disorders. Coccaro and colleagues have found a correlation between mutations in the gene for the enzyme that assists in making serotonin and high levels of impulsive aggression; some of the strongest correlations between the mutation and aggressive behavior came from cases in which an individual had killed a sexual partner. The researchers termed this "outwardly directed aggression," as opposed to "inwardly directed aggression," which is carried to its most extreme form in suicide. As multiple studies have found, diminished serotonin-related activity in the human brain predicts an increased risk for suicide.

Conversely, more serotonin generally means less aggression. For example, Berend Olivier, from Utrecht University in the Netherlands, has charted numerous examples showing that drugs increasing the activity of serotonin will reduce aggression.

Another link between serotonin-related activity and reduced aggression falls into an extremely important domain, the genesis of violent behavior by young men, and brilliantly reveals an example of significant interplay between genes and environment. Avshalom Caspi, Terrie Moffitt, and their colleagues in the Institute of Psychiatry at King's College London, studied the gene coding for the enzyme called monoamine oxidase A, which is responsible for breaking down the serotonin molecule within the tiny space between nerve cells and thereby temporarily putting a stop to serotonin-related activity. They focused on this gene, which is carried on the X chromosome, because previous research in mice and humans had indicated that its absence causes antisocial behaviors.

The King's College researchers wanted to know how different levels of this gene's activity in young men might be connected to the damaging consequences of having been maltreated as children. A so-called short form of the monoamine oxidase A gene produces an enzyme that has an abnormally low level of activity, which means that serotonin, too, cannot exert its effects fully. It looks as though this form of the monoamine oxidase A gene, which was found in about 34 percent of the study subjects, predisposes the person toward high levels of impulsive aggression. Young men with this gene version had higher levels of conduct disorder, a greater disposition toward violence, a higher incidence of antisocial personality disorder, and more convictions for violent criminal offenses than the young men with the regular, long version of the gene. A crucial observation, however, is that these convincing results appeared *only* among young men who had been severely maltreated as children. Thus, two forces for violence, early maltreatment and a genetic alteration, multiplied each other's effects on antisocial behavior. If this exciting finding is replicated, we will have yet another way to explain why violent behavior occurs, when it occurs.

Gene studies in humans underscore the complexity of serotonin's effects. In research aimed at revealing the genetic roots of aggression, Klaus-Peter Lesch, at the University of Würzburg in Germany, has done extensive work to investigate which genes are involved in transmitting serotonin's chemical message in the brain. He discovered that impulsivity and aggression in human volunteers could be linked to increased or decreased activity of numerous genes involved in not just one, but several stages of serotonin's action: in addition to several serotonin receptor genes, they included genes whose job it is to dismantle serotonin or, after the neurotransmitter had produced its effect, move it back into the neuron that had released it. Lesch's findings bring to the fore the difficulty of treating aggression, due to the enormous challenge of developing drugs that would need to interfere with this exceedingly complex system.

The complexity is compounded by emotional factors accompanying aggression. Stephen Manuck and his coworkers have explored a key concept in the neuroscience of aggression: "impulsivity." In contrast to slowly developing, calculated aggression, impulsive aggression bursts out suddenly and is often accompanied by expression of emotions such as anger and rage. Is this outburst accompanied by real feeling, or is it just a reflex? Manuck takes the view that serotonin-dependent effects on aggression are indeed linked to these emotions, and that drugs reducing impulsive aggression are probably effective in part because they also reduce negative emotions in general.

Chemicals of rapid attack

New research suggests that two ancient neurotransmitters with opposing effects—glutamate, which conveys neuronal signals over a distance, and GABA, which blocks these signals—may be involved in controlling aggression. The possible involvement of such old-timers as glutamate and GABA tells us how basic the brain's capacity for aggression really is. These transmitters might take part in governing rapid attack, a kind of aggression that must have appeared early in evolution as an essential means of survival.

Glutamate, one of the most abundant neurotransmitters in the brain, appears to control aggression by acting within the hypothalamus, that multifaceted structure located just above the brain stem that we have seen involved in fear and loving responses. In a part of the hypothalamus whose electrical stimulation rapidly evokes attack behavior, Erik Hrabovszky and his colleagues in the Institute of Experimental Medicine in Budapest, Hungary, found an overwhelming preponderance of glutamate-making neurons. Interestingly, a single enzyme converts this transmitter into GABA—a simple reaction that suggests that the roles of these two ancient transmitters may be closely linked. In fact, Leslie Henderson and Ann Clark, at Dartmouth Medical School, have shown that GABA may play a role in regulating aggressive behavior dependent on male hormones, but how GABA and glutamate interact in controlling aggression remains to be discovered.

Nitric oxide

The important chemical messenger nitric oxide is probably best known thanks to Viagra, the drug that enhances its action in order to stimulate sexual function. But nitric oxide, which is actually a gas, plays numerous and versatile roles in the body, and one of them might have to do with regulating aggression. Randy Nelson, at Ohio State University, discovered by accident that male mice lacking a gene responsible for the synthesis of nitric oxide—and therefore, lacking nitric oxide in their bodies—were highly aggressive toward one another. Through a long and systematic series of studies, Nelson and his students used a variety of genetic and pharmacological tricks to prove that nitric oxide must be thought of as reducing aggression.

Another fascinating finding from the Nelson lab is that female mice lacking the same nitric oxide–synthesis gene never showed unusual aggression. Therefore, these scientists found an example in which the effect of a gene on aggression depends on the sex of the animal. Nelson and his team also produced a nice illustration of an interaction between gene and environment, showing that the impact of a particular gene on behavior is not etched in stone and can be modified by the surroundings. The scientists found that the behavior of male mice genetically engineered not to synthesize nitric oxide in their bodies depended on how these mice were raised: when housed together continuously from weaning until testing, the males revealed a significantly lower level of aggression compared with that of males housed in isolation.

Genetic influences

We know that tendencies toward aggressive behavior can be inherited, and animal breeders have taken advantage of this feature to produce various animals, from mice and rats to fish, birds, dogs, and even horses, with different degrees of combativeness. The genetics of aggression is a field of considerable future importance for human behavior too because studies have shown a strongly inherited tendency toward aggression in humans. For example, studies of twins, which help distinguish between inherited and environmental influences on behavior, suggest that genes account for perhaps 50 to 60 percent of the variation among children in their disruptive, obnoxious behaviors. Although the study of genetic influences on aggression and violence is in its early days, science has already identified a number of genes that influence aggression.

After that simple statement, however, things get more complicated. A panoply of genes is involved in aggression, as in any other complex trait, and that warns us that normal, biologically regulated aggression—as opposed to psychopathic violence—depends on patterns of gene activity orchestrated over time, not on a single gene.

For example, solid banks of data about the genes for estrogen receptors and aggressive behaviors in mice tell us that the relationship of genes to behavior varies depending on a variety of factors. For one thing, my lab has reported that an individual gene can have opposite effects on aggressive behaviors depending on the sex of the animal. Consider the gene coding for estrogen receptor–alpha: when we deleted—or, in technical language, "knocked out"—this gene in *male* mice, aggression plummeted, but when we deleted it in *female* mice, aggressive behaviors increased. For another thing, a closely related gene, coding for estrogen receptor–beta, has the opposite effect: young male mice in which the estrogen receptor–beta gene has been "knocked out" are actually more aggressive than their unaltered littermates in a control group. Third, in female mice, the estrogen receptor–beta gene has different effects on different types of aggression tested. When we "knocked out" this gene, maternal aggression that female mice generally display to protect their pups increased, but aggression unrelated to motherhood decreased.

Knowing that males initiate a large preponderance of violent acts, Steven Maxson and his colleagues at the University of Connecticut analyzed the Y chromosome, the one that accounts for "maleness." In a study of male mice, the researchers identified a particular region of the Y chromosome as especially important for fostering aggressive behavior. From Maxson's work, it is obvious that at least two types of genetic contributions from the Y chromosome are important. First, a gene known by the abbreviation SRY starts a cascade of biochemical reactions that result in the development of masculine genitalia. These in turn produce testosterone, which drives certain gene activity in the brain and, as we have seen above, fosters male aggressive behaviors. Second, other genes on the Y chromosome promote aggression by means that are not yet understood.

Moreover, Robin Lovell-Badge, at the National Institute for Medical Research in London, discovered the gene on the Y chromosome that sets off the chemical cascade leading to male sex differentiation, the developmental process determining that the fetus is going to be male. This differentiation includes, of course, the development of the testes, which, as mentioned above, produce testosterone. Thus, the same genetic mechanism that makes a male a male, automatically sets the stage for aggressive behavior. In short, even without the statistics for men's predominant involvement in violent crimes, all these contributions from the Y chromosome would tell us to expect males to be, on average, more aggressive than females.

To conclude this brief review of genetic influences, let me make a few comments on additional complexities involved. First of all, we must consider that an important factor in allowing a mouse or a man to be aggressive is his level of brain arousal, the general state of the central nervous system that forms the "background" for all our actions, determining their scale and intensity. In my professional text for researchers, *Brain Arousal and Information Theory,* I show that more than 100 genes are involved in the control of arousal; many of these will certainly be found to affect aggression. Furthermore, in this discussion, I have referred to aggressive behaviors as though they were all more or less equivalent to one another. In real life, though, different kinds of aggressive behaviors—territorial attacks, predatory aggression, emotional explosions, defensive behaviors, maternal aggression, and so forth—are likely to bring in the influences of different genes we have not yet considered. Indeed, Edward Brodkin, at the University of Pennsylvania Medical School, has already used a sophisticated genetic technique to show that as-yet-unidentified genes will be discovered that prove to affect aggression in male mice. And finally, still to be brought into the picture soon in this book, is the evidence that connects genes with different temperaments in humans, some of which could bear on propensities toward aggression and violence.

Finding aggression in the brain

The above survey of chemicals and genetic influences brings us to a major issue that could be summed up by one key term borrowed from real estate: location. Identifying the brain regions that promote or, to the contrary, reduce aggression can help us understand how aggressive behaviors come about.

In promoting aggressive behaviors, an important structure for the onset of aggression is the amygdala, the part of the forebrain that also regulates the fear response. While sending out its behavior-affecting signals, this structure is guided by a wide variety of emotionally charged stimuli: for example, a signal stimulating aggression may be triggered by an emotion such as fear. In the male brain, stimulating the amygdala—for example, via electrical stimulation of certain portions of it—incites aggression, while damaging this brain structure does just the opposite: the damage reduces aggression. In both animals and humans, damage to some specific parts of the amygdala was associated with a "taming" effect on behavior.

Pinpointing the precise cell groups in which aggressive behaviors are produced can be important for obtaining the general picture. In the amygdala and another aggression-promoting brain region, the stria terminalis (an emotion-related sinuously shaped group of nerve cells sometimes called the "extended amygdala"), I would put my money on the steroid-sensitive vasopressin-producing neurons as the cellular culprits behind aggression. Practically all of these neurons have receptors for the male hormone testosterone; testosterone stimulates the production of vasopressin, which in turn, as we have seen, robustly stimulates aggressive behavior in a wide range of species.

Apart from the amygdala, vasopressin-producing neurons are present in the hypothalamus, which suggests that this brain structure, which has a day job regulating such functions as the heartbeat and body temperature, is also available to help out in aggression. Other parts of the hypothalamus, not related to vasopressin production, are important to enabling various aggressive behaviors: in cats, electrical stimulation of a particular region of the hypothalamus produces an emotional hissing and biting attack, whereas electrical stimulation of another region produces a quiet, stalking attack.

An additional brain region thought to be involved in promoting aggression is the central gray, a still-to-be understood cell group

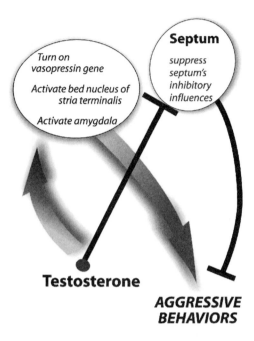

Septum

Turn on vasopressin gene

Activate bed nucleus of stria terminalis

Activate amygdala

suppress septum's inhibitory influences

Testosterone

AGGRESSIVE BEHAVIORS

Figure 9. A two-pronged strategy. Testosterone encourages aggression by both increasing activity [⟶] in nerve cell groups that produce the behavior and reducing activity [━▌] in nerve cell groups that block it.

known to be related to pain. This small, primitive part of the midbrain has ties to our body responses, notably blood pressure, in both fear and sexual behavior, but it also turns out to contribute to aggressive behavior: parts of the forebrain (including, in particular, the hypothalamus) important for regulating aggression send some of their nerve cell extensions to the central gray. Electrical stimulation there causes rage-like responses in animals, which may be the same as "septal rage," a condition described earlier in this book in connection with fear, in which the animal immediately and viciously attacks anything in sight. This may be, in part, because neurons in the central gray also signal pain, which can serve as a trigger for aggressive behavior. In fact, pain, uncomfortable temperatures, social stresses, and other unpleasant environmental influences, according to Leonard Berkowitz of the University of Wisconsin, represent "environmental stimulation to affective aggression." Thus, we can also infer that the sensory pathways carrying those signals join in the neural regulation of aggression, although clearly they carry out other functions as well.

As to inhibiting aggression, the two most potent areas for damping it down are interesting as a team (if a team is what they are): one is a very old part of the brain—the septum—and the other, our much younger prefrontal cortex. It is tempting to imagine that primitive man's reasons to chill out were usually obvious and straightforward, but that, in evolving, we created innumerable scenarios that cast ambiguity on whether to aggress and obliged us to stop and think.

The septum, a small but very important region in the forebrain, powerfully inhibits aggressive behavior, as shown by the fact that damaging this region in an experimental animal can bring on "septal rage." The last time I put my hand in a cage holding a rat whose septum had been damaged, the animal bit me so badly the swelling did not disappear for days.

As for the prefrontal cortex—yes, the same structure we considered earlier in this book as a candidate for inhibiting the output of the amygdala and thus reducing fear response—it seems to reduce aggression through the action of serotonin, the aggression-dampening neurotransmitter. Richard Davidson and his group at the University of Wisconsin believe that the mechanism behind the reduction has to do with the projections of serotonin-carrying neural pathways into the prefrontal cortex. Dysfunctions in these pathways could, hypothetically, explain the behavior of certain violent individuals. Importantly, Davidson has reported that this kind of regulatory neural circuitry "is dramatically shaped by early social influences;" his data and reasoning strongly support the significance of early intervention for the prevention of violence, to be discussed in the next chapter.

If, however, we were to "map out" all the above-mentioned brain regions promoting or reducing aggression, the map wouldn't come out black-and-white: some regions seem to include "islands" of nerve cells that do just the opposite of all the other cells in the same area. To complicate matters further, some brain chemicals seem to exert opposing effects depending on the brain region in which they act.

Thus, within the aggression-promoting amygdala, different groups of neurons are responsible for different effects, and some of these are actually presumed to inhibit aggressive behavior. Take, for example, a portion of the amygdala called the central nucleus. By manipulating this brain region, scientists were able to produce changes in maternal aggression, defensive aggression, and predatory attack in experimental animals; moreover, in humans, this region is involved in the regulation of both reactive and proactive aggression. However, the central nucleus has a large number of neurons that produce the inhibitory neurotransmitter GABA, the rapid-attack chemical mentioned earlier in this chapter. GABA might play a role in reducing aggression because an increased number of its receptors decreases activity in the part of the midbrain thought to stimulate aggression; moreover, at least in some studies, giving GABA or GABA-like drugs decreases aggressive behavior in experimental animals.

Similarly, in the septum, despite its ability to powerfully inhibit aggression, some cells help with fostering aggression. Which ones they are is unclear, but my guess would be the GABA-producing neurons of the region called the lateral septum. These neurons seem to be chock-full of receptors for the male steroid hormones, which, as we know from leading scientists in this field, such as de Vries at the University of Massachusetts, promote aggressive behaviors by acting not only on parts of the brain traditionally involved in stimulating aggression—the amygdala and the stria terminalis—but also on the lateral septum. Moreover, injections of vasopressin into the lateral septum, which almost certainly activates these GABA-producing neurons, stimulate aggressive behavior. Although GABA, as an inhibitory brain chemical, does not sound like a neurotransmitter that would normally increase aggressive behavior, there are several reports that suggest it can do this in the lateral septum.

All the findings reviewed in this chapter—the examples of brain chemicals, genes, and cell groups participating in the regulation of aggressive behaviors—represent just the beginning of intensive study in this area. I said before, but it bears repeating, that aggression has a varied repertoire and a great spectrum. It can range from the merely upsetting reversal of the Golden Rule—a common thing such as stepping on everyone to climb the career ladder—to outrageous and horrifying violations of everything we cherish. There may be common themes amid the diversity of these unethical behaviors, but the more extreme the violent act, the greater the challenge in explaining the urge that people occasionally feel to throw basic decency out the window.

9

MURDER AND OTHER MAYHEM

Family violence. Sexual abuse. Gang fighting. War. Genocide. These are but a few examples of the endless ways in which aggressive behaviors can turn violent. In the preceding chapter, we saw that aggression can be viewed as being embedded in our brains due to the survival edge it provides. Yet the kind of neuroscientific research I conduct deals only with aggression, not violence. Why and how does aggression get out of hand? How do we explain violence?

As I already stated while discussing aggression, explaining the different forms of violence requires a combined scholarly effort by experts from different fields. Individual crime might best be explained by ideas like the "broken windows" theory of criminologist James Q. Wilson. According to this school of thought, when people see small disruptions in the safety and orderliness of their environment—such as a building that appears neglected because its broken windows remain unrepaired for a long period of time—they feel more free to violate larger societal prohibitions, committing robbery or assault, for example. Gang fighting in large cities comes under the purview of sociologists; large-scale warfare, of historians and economists. I cannot pretend even to scratch the surface of these domains of scholarship in this book. I claim, however, that none of these other domains *exclude* a neuroscientific approach to understanding the occurrence of violence, individual and organized. In particular, my theory, though arguing that our brains are wired for ethical behavior, can also offer insights into the origins of violence and perhaps even help develop ways of reducing its occurrence in our society.

Societal roots of violence

A good framework for a discussion of violent behavior is the National Research Council report *Understanding and Preventing Violence.* Though published in 1993–1994, this four-volume document is still the most comprehensive reference work on the topic. According to the report, the frequency and intensity of violent incidents vary greatly across periods of time and geographical regions; for example, countries with the highest violence rates, such as the United States, have ten times more violent incidents compared with places with the lowest rates, such as England or Singapore.

Beyond the statistics, however, the National Research Council report dwells on the causes of violence. Among these are causes that primarily have to do with the situation of the individual. For example, the rate of violent crimes per capita is higher in large cities than in small towns—apparently due to socioeconomic factors, such as the humiliations of poverty, and the greater opportunity for crimes against persons in the crowded surroundings of a city. The report also draws on a psychological theory of John Dollard and Neal Miller, which states that violent behaviors can take the form of learned responses to frustration. This theory is based on findings from experimental laboratory settings, in which animals and humans have been shown to become very aggressive when their paths to highly desired incentives are blocked.

What might be the antecedents of violent acts? Risk factors include poverty during childhood, often coupled with low birth weight; trouble during preschool and early school years, often coupled with poor concentration and acting out; bad parenting—in the language of the report, "harsh and erratic discipline, lack of parental nurturance, physical abuse and neglect;" poor supervision during a boy's early family experiences; and all the "factors associated with large low-income families," including residence in high-crime neighborhoods. To make matters worse, problems related to alcohol and drug abuse are compounded by the availability and use of firearms. In the United States, readily available guns both multiply the opportunities for damaging, aggressive behaviors and greatly magnify the consequences of those behaviors.

As for the organized forms of violence—such as gang fighting, terrorism, wars, and attempts at genocide—their psychological and socioeconomic causes are also varied. Sociology and psychology have the most to say on these causes at this time, but neuroscience is gearing up to contribute: the brain is built for social interaction, and researchers are learning to design studies intended to reveal the brain's actions in group situations.

For example, the social psychologists Joel Wallman and Karen Colvard, whose work at the Guggenheim Foundation centers on the causes and reduction of aggression, say that the emotional underpinnings of different forms of violence surely are different from one another. Their studies suggest to me that in some of the most highly organized aggressive groups, the violence committed by an individual may actually reflect some kind of bonding with other members of the group. And bonding is a brain question; in fact, it's an issue that—paradoxically enough, considering the violent context of organized crime—links right up with the brain's Golden Rule mechanisms of shared identity proposed by my theory.

Loyalty as well should be accessible to brain research. We should be able to ask what happens in the brain of a gang member when he commits a violent act because he does not want to let down his comrades. In gangs, aggressive acts might be both an issue of loyalty and a requirement for membership. In the words of psychologist Clark McCauley and political scientist Brendan O'Leary, of the Solomon Asch Center at the University of Pennsylvania, "Humans are true to their groups." If loyalty to the group requires violent acts, group members will perform violent acts—especially if they are young men. McCauley considers that even some terrorists might be seen as normal people responding to threats against their cause or their group.

Joining with John Hagedorn of the University of Illinois at Chicago, I picture the attractiveness of gang membership as depending on the insecurities of powerless young men who have suffered humiliation in society and are seeking to exercise their will in hatred of women and aggression in general. However, the formation and durability of gangs resist easy analysis. In Hagedorn's view, all the social and economic forces that were involved in the industrialization and urbanization of cities during the twentieth century have played a part: poverty, racism, oppression, and the consequent humiliation of young men.

In explaining the causes of violence, neuroscience can perhaps contribute the most at the interface between societal phenomena and insights into the individual brain. One particularly illuminating area of research in this respect is the study of substances that temporarily alter people's behavior.

Are the Crips teetotalers? Or the Bloods, or any unit of the military heading out on weekend leave? When aggression escalates into violence, one of the causes is alcohol consumption. Epidemiological studies convincingly link alcohol to aggression. In the laboratory

setting, as well, controlled studies confirm that aggressive responses increase after moderate or high doses of alcohol. Furthermore,
people addicted to any of a variety of illegal drugs may turn to violent behavior in order to gather resources to buy those drugs. Neuroscience has decades of findings on the brain's response and ad-
aptation to alcohol and addictive drugs, and these, in turn, might
contribute to the understanding of violence.

Thus, Klaus Miczek at Tufts University, an authority on alcohol's effects on brain and behavior, has looked for the differences
between the brains of people who become more aggressive while
drinking and those who do not. He and colleagues found that these
two groups of people differed, among other features, in the genes
manufacturing the receptor for GABA, the inhibitory neurotransmitter that we saw in the preceding chapter to be linked to aggression. Findings of this sort may help explain why certain people are
prone to become violent after consuming alcohol, and perhaps in
the future even facilitate the development of preventive measures.

Another social phenomenon that has deservedly attracted the
neuroscientists' attention is the popularity of synthetic male hormones, the anabolic androgenic steroids, which have more grim effects on the brain. Their use has escalated, especially in teenagers,
in some cases to enhance athletic performance and in other cases
to look good—that is, to develop bigger muscles and improve one's
body image. These drugs can cause the so-called 'roid rage, a sudden explosion into violent behavior. Worse, their potential for long-
term repercussions on aggression is unknown.

Marilyn McGinnis, now at the University of Texas at San Antonio, has investigated the behavioral effects of these drugs in laboratory animals. In a study using mice to compare the effects of several steroids under different experimental conditions, she found that
the hormones' impact on aggressive behavior depends upon which
steroid is being used (some increase aggression while others actually decrease aggression), and also upon the provocation by environ-

mental cues (while defending their home cage and in response to pain, hormone-loaded males were more aggressive when confronted with other hormone-loaded males). McGinnis feels that male hormones do not simply drive aggression, but that, in her words, "anabolic androgenic steroids sensitize animals to their surroundings and lower the threshold to respond to provocation with aggression." Since she studied pubertal male animals, her results may be relevant to the troubling incidence of violent acts among adolescent boys and young men, particularly since these young people, in addition to possibly taking synthetic hormones, might have their own high testosterone levels, which some researchers associate with antisocial personality characteristics.

In animals and humans, however, not all synthetic male hormones affect aggression. Sonoko Ogawa, working with me several years ago, studied a synthetic steroid called methyl-nortestosterone, which has testosterone-like properties and cannot be converted into other steroid hormones. Injecting this steroid into castrated male mice and comparing its effects with those of testosterone, we found a substantial difference: the synthetic hormone restored male sexual behaviors without stimulating aggressive behaviors, while testosterone could bring both sex behaviors and aggression back from castrated levels to normal levels. This finding does not refute the above-mentioned behavioral risks of synthetic male hormones, but it does point, yet again, to the complexity of the hormonal and other mechanisms regulating aggressive and violent behavior.

Biological origins of violence

In enumerating the biological factors contributing to violence, the National Research Council report basically follows the mechanisms—biochemical, genetic, and neuronal—that I described for aggression in the preceding chapter. It also states that violence can sometimes be conceived as a product of brain damage (as in violent children who have brain-damage-related "neuropsychological deficits in memory, attention and language/verbal skills") and covers various contexts of unacceptable aggression, including that of mental disorders.

In discussing aggression, I have described it as a more male than female phenomenon. I have also mentioned that arrest rates for murders, rapes, robberies, and aggravated assault across lifetimes all closely follow the lifetime curve measuring testosterone in the bloodstream. The testosterone connection appears to hold not just for normal men but also for those with personality disorders, and not just for crime but also for a trait defined as "sensation-seeking," which can easily lead to reckless behavior. For example, the National Research Council report cites the studies conducted by Emil Coccaro, at the University of Chicago. Besides observing correlations between the bloodstream concentration of testosterone and aggression in normal men, he found a correlation between testosterone in the cerebrospinal fluid and sensation-seeking among men with personality disorders.

Other findings by Coccaro provide support for the identification of brain regions we have seen to be involved in suppressing aggression, particularly the prefrontal cortex. He has studied a type of pathological behavior he calls "intermittent explosive disorder," an extreme form of impulsive aggression. He found that this disorder developed early in life, especially in male patients, and that these patients tended to perform on psychological tests in ways that revealed damage to the prefrontal cortex. Likewise, Adrian Raine, at the University of Southern California, who conducted magnetic resonance imaging of the brains of men with antisocial personality disorder, reports the amount of prefrontal cerebral cortical tissue in their brains to be significantly below normal.

This lack of suppressive mechanisms also emerges in studies of teenage boys regularly engaging in aggressive behaviors. In Richard Tremblay's research, launched with his team at the University of Montreal, the biggest red flag in early childhood of such boys was not the mere appearance of physical aggression, but the inability to inhibit it by normal social cues. The mothers of these boys had been unable to teach them normal social inhibitions; the sons, already physically aggressive at the age of six, continued to be so into their teens. Study authors have also found that violent teenagers might be described as having flat, deadened emotional responses to the problems of others, following a personal history of neglect and humiliation.

Research indeed reveals a connection between emotional responses and a brain structure crucially involved in regulating aggression, the amygdala. Peter Lang and his team at the University of Florida, building on results with experimental animals by Michael Davis, at Emory University, obtained insights into this matter from the responses of human subjects to three different types of pictures. Two of the types were highly arousing emotionally: either very pleasant (babies, puppies) or very unpleasant (guns, snakes poised to strike, mutilated bodies). The third type was emotionally neutral (baskets, trees). When extremely unpleasant pictures are presented to normal, nonviolent people, they cause a startle reflex and activation of the amygdala. Damage to the right amygdala inflicted by accident or disease blocks this reaction. And in fact, people who have been diagnosed with psychological abnormalities exhibit a characteristic deadened response to emotionally arousing stimuli that suggests their amygdala is functioning abnormally.

Reducing violence in our society

Do we know enough to prevent or reduce violent aggression, or at least its most worrisome forms? Right now, early in the twenty-first century, there is no way that neuroscience can simply wrap up the problems of violent behaviors with a new technique or two. Faced with the grim list that began this chapter, a variety of experts—economists, sociologists, historians, psychologists, and more—all have something to say about the motives, normal or abnormal, that lead to violence. Among other observations, as mentioned earlier, they point to the disturbingly wide variety of aggressive behaviors. Clearly, in the future, different kinds of violence, each with its distinct pattern of causation, will yield to different strategies for reduction.

The report of the National Research Council suggests taking into account both the societal and the biological causes of violence, including those listed in this chapter. It recommends paying special attention to predispositions toward violence based on genetics, early childhood mistreatment, and possible brain damage; thus, resources directed to minimizing these factors would be time and money well spent. The authors of the report also recognize a special susceptibility in boys at the time of puberty, when the appeal of socialization (or antisocialization) into a gang may be amplified by sharply rising testosterone levels. Accordingly, making allowances for the normal effects of testosterone on a boy's physical activity and providing meaningful and nurturant alternatives to gangs would be particularly helpful.

In other forms of violent behaviors, altering the circumstances might help avert trouble. With respect to violence connected to commercial robberies, the report recommends "modifying places, routine activities and situations that promote violence." This commonsense recommendation could also apply to high-risk situations for sexual and school violence.

As mentioned above, a vast literature exists to show that alcohol intoxication encourages aggressive behavior, even to the point of criminal violence. Tufts University's Miczek has written that more than half of damaging violent behaviors are connected with alcohol abuse. Therefore, according to the report, groups such as Alcoholics Anonymous can make an important contribution to reducing violence in our society—although Miczek is quick to add that "the link among alcohol, other psychoactive drugs and violence turns out to be not an example of straightforward causation, but rather a network of interacting processes and feedback loops."

Prison sentences for violent behavior seem like a logical way to deter further violence, but this effect is surprisingly hard to prove. Although the report reminds us of the possibility that "incarceration may reduce violent crime through two mechanisms, deterrence and incapacitation," James Gilligan, former chief of psychiatry for the Commonwealth of Massachusetts and now visiting professor of psychiatry and social policy at the University of Pennsylvania, has been quoted as saying that "prisons make violent persons still more violent." It looks to me as though the evidence of a deterrent effect is actually rather weak.

Summing up these recommendations, I must admit that though the National Research Council report is an immensely valuable document, I found its suggestions disappointing. Instead of a crisp, logical, well-organized, and intellectually powerful set of recommendations based on solid biological, psychological, and sociological literatures, it simply called for further research. We know that new and practical actions are needed; the incidence of aggressive acts, individual and social, in our country and around the world, proves it. What we need is a compelling and structured plan of action.

Preventive measures

What are the ways in which at least some of these forms of violence may be prevented? To answer this question, I look to the wisdom and experience of specialists like Colvard and Wallman, of the Guggenheim Foundation. Their message is sobering: at this point, any discussion about the prevention of violence has to be short and modest because, basically, we are not very good at preventing violence.

The search for more effective strategies for the prevention of gang violence, to consider just one example, takes in a broad range of scholarly fields. A multidisciplinary scientific approach to the prevention of youth violence has been developed by John Devine and his colleagues on the basis of a public health model initiated by James Gilligan in his book *Preventing Violence.* This approach sets out three different levels of prevention.

The measures that come under the heading of "primary prevention" apply to the population as a whole, regardless of each person's current health or risk of illness in the future. With respect to youth violence specifically, Devine and his colleagues offer four recommendations: (i) reduce extreme socioeconomic disparities and try to reduce the effects of such disparities; (ii) encourage smaller school sizes; (iii) promote rituals (like initiation rites) that offer young men and women positive visions of their adult roles in society; and (iv) reduce the consequences of impulsive aggressive behaviors.

The next level, "secondary prevention," calls for programs to help young people who carry an "increased risk of [committing] violence, even though they have not yet committed any acts of violence." In addition, according to the Devine group, professionals should examine and react to potential predispositions toward violence; improve the care of children during critical periods of high vulnerability; work toward the reduction of bullying; and insist upon the avoidance of alcohol.

"Tertiary prevention" refers to helping "individuals who have already demonstrated violent . . . behavior" in order to reduce the probability of its recurrence, or recidivism. Other helpful measures would include avoiding punishments that are counterproductive, considering the role of testosterone in aggression, and trying to intervene with drug treatment.

I note that while these lines of thinking were proposed to deal especially with young males, many of them could apply throughout much of life. One thing that has become clear from many studies is that the earliest interventions are the most effective. In particular, gaining the allegiance of the child before he joins an antisocial peer group is most useful, according to the studies of Delbert Elliott at the University of Colorado. Even better than reducing the rate of gang violence would be reducing the rate of recruitment into the gang in the first place.

The call for intervention as early as possible appears quite realistic because, as Rolf Loeber at the University of Pittsburgh says, "in the majority of the most seriously violent cases, the behavioral problems date back to the early childhood years." They are detectable, and therefore they can be confronted. Sylvana Cote and Tremblay, in Montreal, detected stable and consistent patterns of high or low physical aggression in children from the ages of two to eleven. It appears to me that as these children learn properly (or fail to learn) to inhibit physical aggression, feedback loops may operate in either direction, forming either vicious circles or splendid circles. In other words, if there are problems between two little boys, their failing to learn social restraint leads to fights, which will lead to more fights; but if both the boys learn prosocial behavior, they will reinforce each other's avoidance of antisocial, combative acts.

Violence against women

Of all the attempts to bring inappropriate violence under control, I am most concerned about the woeful state of efforts to reduce violence against women. This problem has been documented in virtually every country of the world. Enrique Echeburúa, from the University of the Basque Country, in Spain, together with researchers at the Open University of Madrid, has looked for abnormal psychological characteristics in 54 men who were in prison because they had committed serious violence against girls or women. Divorced men and widowers far outnumbered married men. Their personality tests did not show anything remarkable, according to Echeburúa. While they tended to be depressed and anxious, with paranoid thoughts, they did not have a high level of readily recognizable psychological symptoms. From the point of view of a scientist searching for violence reduction strategies, this relative lack of conspicuous psychological characteristics is disturbing, because it gainsays therapeutic approaches that have been worked out for other behavioral abnormalities.

Back in the United States, according to the report of the National Research Council, the few school-based programs that have been initiated to prevent violence against women have not been thoroughly evaluated, and deterrence through the criminal justice system obviously has not worked. The assessment of therapeutic programs for "batterers' groups" has often relied on information from the batterers themselves. However, those that have relied on information from the partners of the batterers have reported better-than-halfway success in periods of up to two years, but with rates that range widely, from 53 to 85 percent. As for couples therapy, it has failed to lower physical aggression significantly.

In fact, the situation regarding violence against women looks so gloomy just now that it may be appropriate to consider quite drastic measures. For example, one strategy for combating such violence might rely on the fact that, universally, sex behavior by men depends on testosterone circulating in the blood and binding to testosterone receptors in the brain. It turns out that prolonged administration of a synthetic version of the brain chemical called luteinizing hormone-releasing hormone, abbreviated as LHRH, turns off the pituitary gland machinery for stimulating testosterone production by the testes and lowers libido.

What about abnormal sexual behavior itself? Justine Schober, an academic medical doctor with long experience practicing in Erie, Pennsylvania, had great sagacity in choosing this LHRH therapy in the treatment of pedophiles, men whose sexual urges and/or fantasies involve children. The treatment significantly reduced sexual urges toward children. Her findings resonate not only with those of another expert, Fred Berlin, at Johns Hopkins School of Medicine, but also with a Canadian report. Cooper and Magnus, working at Saint Thomas Psychiatric Hospital in Ontario, found that a different approach to blocking testosterone actions—giving a steroid drug that blocks access to testosterone receptors—reduced sexual arousability.

The study most germane to our discussion is one from Briken and Michl, from a very sophisticated clinic for sexual investigation at the University of Hamburg Medical School in Germany. When they gave the long-acting LHRH treatment to sexual offenders, the recipients' sexual aggressive impulses were eliminated almost entirely. In cases in which violence against women partakes of this kind of sexual aggressiveness, then the long-acting LHRH treatment is worth trying. Sex offenders can also be treated with Lupron, a synthetic form of LHRH, a medicine that works on the pituitary gland to shut down production of testosterone in the testes. One major problem with all these therapies, however, is that many offenders refuse treatment.

Staving off trouble

With respect to criminals, John Laub from the Department of Criminology at the University of Maryland, has begun to describe some of the factors in the lives of men who have turned away from crime. He finds that an increased sense of "structure in their lives" seems to help; this could come from being in the army, for example, or from getting married or holding down a job. Laub also writes, more vaguely, of the need for a "knifing off" (that is, making a clean break) of a young man's personal conception of the past from his conception of the future. Psychologically, as I understand his point, the criminal's past must be separated from future-oriented commitments, directions, and new sources of meaning in life. A guy needs a fresh start. The Chicago researcher Hagedorn wants to "bring gangs and the underclass into the polity," incorporating their members into broader social movements that are not violent.

Sympathetic to this point of view is sociologist Robert Sampson, of the University of Chicago, whose study stresses, instead of external actions such as a police crackdown, "the effectiveness of 'informal' mechanisms by which residents themselves achieve public order." Sampson cites "shared willingness to monitor children's play groups, help neighbors and intervene in preventing acts such as juvenile truancy or street corner loitering [as] key examples of neighborhood collective efficacy." On a similar note, Felton Earls, at the Harvard School of Public Health, emphasizes the importance of residential stability as an overlooked feature in maintaining safe neighborhoods. All these examples and approaches have in common the inestimable feature of reducing violence by nonviolent means.

In seeking to avert terrorism or politically motivated violence, it seems important not to escalate negative feelings. McCauley, from the Solomon Asch Center, says states should use nonviolent means, including the criminal court system and ways of reducing terrorists' means of support, financial and otherwise. Taking all possible measures to "reduce the flow of new sympathizers," according to McCauley, will help to solve the problem of terrorism without the use of military force.

Can we sometimes nip potential problems in the bud? Is it possible, at least in some cases, to intervene in children's early lives in such a way as to prevent serious antisocial behavior? Kenneth Dodge and John Coie, professors of psychology at Duke University, participated in the work of a team called the Conduct Problems Prevention Research Group. They have initiated a program called Fast Track, a ten-year project that involves classroom teaching of emotional concepts and self-control, home visits for the purpose of fostering parents' problem-solving skills, and child social understanding training groups called Friendship Groups. It is too soon to know how effective Fast Track will be in the long run, but a comprehensive data collection center run by the program certainly will let us know its strengths and weaknesses.

Finally, with respect to the medical means of ameliorating aggressive tendencies in violent individuals, it looks to me that our efforts are only in very early stages. In treating violent patients, physicians seem to have no better way of selecting appropriate drugs than by trial and error. Perhaps as we enter the era of pharmaco-genics—during which we will be able to look at a patient's genetic history and predict which medications will be most effective—the reduction of aggression by individuals through medications will be achieved more effectively.

A fuller understanding of the underlying brain mechanisms of aggression and violence will surely help scientists develop more so-phisticated drugs. One example can be drawn from the work of the Dutch researcher Berend Olivier, who went to great effort to develop drugs that would reduce aggressive behavior by affecting serotonin-related neuronal systems. Indeed, he found substances he called "serenics" because the animals given the drugs seemed se-rene. However, these drugs also reduced the overall likelihood that the animals would move or carry out any activity voluntarily. I specu-late that this happened because Olivier's drugs in fact reduced gen-eralized central nervous system arousal, which, as I explain in my book *Brain Arousal and Information Theory,* underlies all expressions of emotion. Since high levels of arousal are necessary for aggressive behavior, its suppression might represent one of the mechanisms by which drugs decrease aggression. This thinking lays bare a major challenge: how do we reduce aggression without reducing the gen-eralized central nervous system arousal that powers the expression of all our emotions? The challenge is likely to produce a great deal of interesting research in future years.

In short, there is no simple "technical fix," no single solution to the problems mentioned in this chapter. However, as an optimist by nature, I believe that civilized expressions of shared fears and shared hopes must reside deep in our brains' mechanisms quite as powerfully as the violent inclinations surveyed above. Otherwise, human society would have fallen apart a long time ago.

10

BALANCING ACT

My neuroscientific theory of fair play has brought us to an interesting juncture. On one hand we have a plausible explanation of how we achieve behavior that obeys a universal ethical principle, but on the other hand we have examples of behavior by individuals and by organized groups that flagrantly disobey the same ethical principle. Now let's deal with the contradiction.

In the human central nervous system, the processes that underlie ethical behaviors must be *balanced against* the mechanisms that cause aggressive behaviors. The blurring of identity that we have seen to result in shared fears and in friendly, affiliative tendencies is opposed by aggressive tendencies which, as we know, can get out of hand. What is the "balancing" I am talking about and how is it reached? Why are some people civil and naturally friendly whereas others are more prone to show hostility? And what about all the social behaviors and responses in between? I believe that the balance depends crucially on a person's temperament. In turn, temperament depends on the complex mixture and interaction—unique for each individual—of genetic influences and environmental forces.

Lifelong traits and temperament

For the past hundred years, psychologists, psychiatrists, and behavioral scientists have been gathering evidence that some of our personality characteristics remain stable through virtually our entire life. One of the earliest in this line was a founder of psychoanalysis, the Swiss medical doctor Carl Gustav Jung. He derived his theories of personality from observations of his own patients and set them forth in his 1921 book *Psychological Types.* Jung sorted people's personalities according to their cognitive styles. His rather rigid "types" of personality characteristics manifest themselves in four main forms of cognitive activity: thinking, feeling, sensation, and intuition. "If one of these functions habitually prevails," says Jung, "a corresponding type results." In one of his most famous formulations, Jung described some of his patients as "extraverts," who had a "positive relation" to others and who oriented themselves "predominantly by the outward collective norms, the spirit of the times," in other words, the people and society around them. Other patients, the "introverts," had their thinking, feeling, sensations, and intuitions oriented not toward others but by their own inner, subjective world.

Thus, some of us might be thinking people, concerned with intellect and ideas. Others, feeling folks, focus on emotions. Still others, sensing individuals, take in information primarily through our perceptions; finally, some are intuitive. (Modern psychologists have used Jung's observations to construct commercial personality typologies such as the Meyers-Briggs Type Indicator, sometimes used in career counseling and in job interviews, and most recently adapted for use in online dating services.) In some of his patients Jung found these personality characteristics so strong and conspicuous that he assumed, without any further evidence, that they were "in principle, hereditary, and inborn in the subject."

With respect to modern behavioral science, I have been inspired to consider human temperaments in terms of lifelong traits by the writings of Mary Rothbart at the University of Oregon. She defines temperament as "constitutionally based individual differences in re-activity [responsiveness to change] and self regulation in the do-mains of affect, activity, and attention."

A prominent psychologist and academic of the mid-twentieth cen-tury, Gordon Allport, at Harvard College, also emphasized that an in-dividual's behavior could be influenced throughout his life by deter-minants he called "traits" and by temperament, which he regarded as biologically determined. Traits are inferred from a person's behavior and, according to Allport, refer to a "neuropsychic structure" that guides a person's responses to a wide range of stimuli in a mean-ingful and self-consistent manner. One person might generally be described, for example, as "conservative," and this trait is likely to come through not only in his political views and his style of clothing but also in his mode of speech and his courtliness toward the oppo-site sex. A person known as "generous" might perhaps be both un-punctual and forgiving of unpunctuality in others, easily persuaded to share his possessions, and tolerant of viewpoints different from his own. And so on, in somewhat stereotypical terms. To this extent, a person's reactions to certain situations would be predictable.

Later, Calvin S. Hall crystallized these concepts and put them in a textbook form that all students could understand. Explaining the essential idea, Hall wrote, "A person's trait . . . describes a cluster of behavioral responses elicited by a group of stimuli."

A theory developed by psychologist Hans J. Eysenck at the Uni-versity of London's Institute of Psychiatry emphasizes the long-term biological determinants of such traits, including differences be-

Figure 10. Spotting a trait. These pairings illustrate C.S. Hall's idea that a personality trait can be seen in an individual's responses to certain types of stimuli. Here the cluster of responses to various types of people reveals a distrustful personality.

tween individuals in the levels of inhibitory and excitatory relations among neurons, differences in their autonomic nervous systems, and genetic influences that cause individual differences in behavior. Eysenck used statistical, quantitative methods to discern long-lasting personality traits such as "rigid," "controlled," "impulsive," "restless," "outgoing;" his distinctions between extraverts and introverts match Jung's very well.

Jerome Kagan, a professor of psychology at Harvard University, is famous for his insights into child development, which are based on hundreds of interviews with children and their parents as well as thousands of hours of observation. His results emphasize features of personality that are identifiable in young children and that last well into adulthood. In Kagan's view, each individual's unique personality is influenced by inheritance, as are the child's viscera and vital organs, and becomes evident very soon after birth. In his 1998 book, *Three Seductive Ideas,* Kagan wrote: "A moral motive and its attendant emotions are as obvious a product of biological evolution as digestion and respiration."

A key distinction, for Kagan, is between "high reactive" infants, who make vigorous movements and show distress in response to certain environmental stimuli, and "low reactive" infants: infants with a high-reactive, "inhibited" temperament grow into individuals who tend to avoid novelty, while, in contrast, "low reactive," uninhibited infants grow into individuals who welcome novel people and situations. A scared, inhibited child, who might be said by Kagan to have a "reactive" sympathetic autonomic nervous system, is more likely to become worried and to avoid social encounters, whereas a low-reactive child will tend to become a calm, sociable adult.

Kagan has added studies of the brain to his evidence for stable features of personality structure. He and his colleagues carried out functional brain imaging on young adults whom they had observed years earlier as small children. As adults, those whom the scientists had originally labeled "inhibited" now showed greater activation in the amygdala, the brain region we have seen to be most associated with fear, than did those who had been "uninhibited" babies. In another study along the same lines, ten-year-old children who had been "high-reactive infants and fearful toddlers" showed unusual activation in the right prefrontal cortex, a brain region I have discussed in Chapter 4 in association with negative moods and poor feelings of self. In sum, Kagan feels that brain and autonomic nervous system physiology go along with the child's behavioral history in predicting current temperamental status, and that certain traits of this temperament stay with us throughout most of our adult lives.

Adrian Raine concurs. A scientist in the Department of Psychology at the University of Southern California, Raine specializes in linking physiological events throughout the brain and the body to their behavioral consequences. He is best known for linking prefrontal injury in childhood to homicidal behavior, but in the 1990s he established the theory that brain damage in the visual system could impair a newborn's ability to bond with his mother, thereby interfering with early attachment and regulation and raising the baby's later risk of developing persistent antisocial behavior. In a 2005 paper aimed at uncovering the impact of cognitive impairments on behavior, Raine easily found substantial numbers of boys aged 16 to 17 years who had shown seriously antisocial behavior, much greater than that of age-matched controls, during the previous ten years. The acts of delinquency ranged from stealing to gang fighting to car theft, and even to murderous attacks with intention to hurt seriously or kill. Thus, in these boys, this temperamental disposition was found to persist for years.

From these and similar studies, a consensus has begun to form as to the characteristics of human behavior that typify an individual's personality virtually throughout his or her life. Of course, some traits may be deeper or more important to shaping a person's behavior than others. The late British-born psychologist Raymond Bernard Cattell, who developed an influential theory of personality, distinguished surface traits—collections of behavioral tendencies that seem to go together—from source traits that he considered to be "the real structural influences underlying personality." For example, detailed statistical analyses have led to the identification of source traits such as conservative temperament, submissiveness, and artlessness.

While discussing different personality types, we must also consider the "scale" of a person's behavior, the arena in which that person's behavior affects others. For example, the financier Alberto Vilar had a personal passion for philanthropy to cultural organizations. This was wonderful, but as James Stewart wrote in the *New Yorker*, he apparently shows the opposite of generosity in his business dealings: he has been charged with defrauding investors and embezzling several million dollars from an heiress whose accounts he had been hired to manage. Such caveats about the relative importance of various traits or seeming contradictions within one personality need not distract us from our main point, which is that certain characteristics persist from one's earliest years throughout much of life. These stable tendencies may, in some individuals, regularly produce civil, friendly, affiliative behaviors, and in other individuals antisocial, aggressive, and even violent behaviors.

Where do these temperamental characteristics come from? Certainly genes play a role, as we shall see in the rest of this chapter. Even animals have stable, distinct features of behavior that can be bred, thus indicating the influence of heredity. But we shall also see later that the genes for these features do not exert their influence at a steady rate over the years; the course of development often runs through environmental "hot spots."

The genetic origins of personality

When we talk about genetic influences on temperament, it is easiest to start with animals. Not only scientists but most pet owners would readily agree that at least some animals—those with whom we frequently interact, like cats, dogs, and horses—have personalities. It is not altogether surprising to hear about the observations of Brian Hare and his colleagues at the Max Planck Institute in Leipzig, Germany, who proved the existence of cooperative instincts among chimpanzees. The experiment involved pulling on two ropes to get food. Sometimes the ropes were too far apart for any individual to reach both. In cases in which the chimpanzees could benefit from cooperation, if they had a choice between two helpers, they usually chose the better rope-puller.

Some scientists detect personality in lower animals, as well. Jason Watters, of the University of California at Davis, sees it in his favorite insect, the water strider: "We do know that these water striders express consistent behavioral types. In the presence of a predator some individuals will run and get right out of the water. Others don't seem concerned whatsoever—just sit there. Others get out and then get back in after a little while." Likewise, Göran Nilsson and his team, at the Department of Molecular Biosciences at the University of Oslo, Norway, distinguish "personality differences" among fish—or, to be precise, trout. Nilsson observes differences between some trout that react to the presence of other fish by actively "coping" with the situation, and others that respond passively by simply withdrawing to a safe area. As for my own experiments, I have found that individual mice show reliable and persistent differences in their level of brain arousal, as well as in the levels of activity and aggression.

Agricultural scientists were the first to see evidence that genes influence behavioral traits, from breeding animals to minimize or maximize those traits as needed. In animals as various as fish, mice, rats, dogs, and horses, we can influence temperament by breeding them with specific dispositions in mind: consider Siamese fighting fish, or compare pit bulls with fox terriers. We can think, too, of high-strung, elite race horses with the calm companion horses that walk next to them to quiet their overreactions to startling stimuli. Thousands of examples like this indicate, although they do not prove, the contributions of these animals' genes to their typical states of behavior. It is possible, however, to show proof of the genetic contribution, and scientists have done so in cross-fostering studies, those in which babies of one genetic strain are raised by mothers of another genetic strain to separate the genetic variable from the variable of parental experience. In short, it has been well established that animal genetics and neurobiology have much to contribute to our understanding of the origins of temperament.

When medical doctors began to gather evidence for the influence of genes on temperament in humans, they started with very different intentions from those of the agricultural scientists or laboratory researchers mentioned above. What they wanted to discover first were the genes that might predispose a person to extremely different patterns of behavior that we identify as mental illnesses. From this research, we know that a portion of a depressed patient's problem with mood is likely due to a genetic influence. The same is true for schizophrenia: several labs, such as those of Daniel Weinberger, at the National Institute of Mental Health, and Maria Karayiorgou, at Columbia University Medical School, have participated in large multicenter studies to come up with more than ten sites on human chromosomes that are associated with this disabling brain illness, which has been found in about 1 percent of American adults. Genetic changes associated with these major mental illnesses can be subtle. The substitution of just one DNA building block for another can change the code in that part of the DNA such that a different amino acid will be made. And if that amino acid affects the shape of its resultant protein—be it an enzyme, a structural protein, or a substance that the nerve cell will secrete—all hell breaks loose.

From mental illness, scientists have moved on to investigate genes that affect temperament in healthy individuals. Studying these genes in extreme close-up, so to speak, is one way to find out what they do; another way is to compare the behaviors of identical twins with those of fraternal twins. Both types of twins develop in a shared uterus and usually grow up in a shared household, but whereas identical twins, originating from a single fertilized egg, have all their genes in common, fraternal twins—who originate from different eggs that are fertilized at the same time—have, like any other siblings, only about half of their genes in common. While relying on twin studies, a multiuniversity team in the United States, represented by John Alford of Rice University, has estimated that as many as 25 to 50 percent of the factors that determine people's opinions about social issues such as school prayer, X-rated movies, divorce, abortion, and the death penalty can be attributed to genetic influences. Likewise, Dorret Boomsma and his colleagues at Vrije Universiteit in Amsterdam, the Netherlands, used data from twins to suggest that genetic contributions account for 48 percent of the variations in loneliness experienced by adults. Interestingly, this figure is very similar to that reached by previous smaller studies in children estimating the genetic contribution to loneliness to be about 50 percent.

Most pertinent to the topic of this book, twin and adoption studies first told us that tendencies toward violence and other antisocial behaviors are heritable. Since then, several studies have suggested that 30 to 50 percent of the tendency toward aggression in adults can be attributed to genetic factors. In children, according to a review of the literature by Daniel Blonigen and Robert Krueger, from the University of Minnesota, about 60 percent of the influences toward high aggression are due to the genes. These are rough estimates because of differences in scientific methodology from one study to another. But beyond these quantitative measurements lies another, perhaps more fruitful question: which genes might be involved?

Serotonin, dopamine, and "prosocial" chemicals

In these early stages of the investigation, researchers are exploring the genes coding for the brain chemicals that we already know are connected with emotions, personality, and temperament. Genes having to do with the neurotransmitter serotonin have offered an obvious place to start, as serotonin is known to affect mood and temperament. It's of special importance for our discussion that, as we saw earlier, this neurotransmitter also affects aggressive behavior: low levels of serotonin, high aggression.

Important studies of serotonin and of another temperament-setting neurotransmitter, dopamine, have been conducted by C. Robert Cloninger, a professor of psychiatry and genetics at the Washington University School of Medicine, who has provided much leadership in the international effort to discern the biological sources of human temperament, including its genetic roots. In particular, he has connected serotonin and dopamine with several dimensions of temperament that represent important aspects of human behavior. (Dopamine is best known as the neurotransmitter for movement; for example, it is the receptors for dopamine that degenerate in the disorder we know as Parkinson's disease. It is also well known to neuroscientists, however, that certain variations in the molecular structure of dopamine receptor genes have been associated with high levels of aggression.)

First, genetic variations in a serotonin-related region of DNA affect what Cloninger calls a "thrill-seeking" tendency—that is, a bent for impulsive behavior that could be considered spontaneous or reckless, depending on the circumstances. This stretch of DNA exerts controls over the gene that codes for a protein whose job it is to remove serotonin from the synaptic cleft—the tiny space in which communication between nerve cells takes place—and thereby put an end to the influence of serotonin on that brain circuit for the time being. A particular genetic variation that tends to leave lots of serotonin in the synaptic cleft seems to contribute to thrill-seeking. Second, variations in the gene that breaks down serotonin into smaller compounds for later reassembly have shown connections with an aspect of behavior that Cloninger characterizes as "persistence." And third, changes in a particular dopamine receptor gene referred to as D4 are connected with the alteration of an important and well-established feature of temperament called "harm avoidance." Research of other scientists supports and extends Cloninger's findings. Thus, John Fossella, working with cognitive neuropsychologists Michael Posner at the University of Oregon and James Swanson, University of California at Irvine, found that different variations in this dopamine receptor gene are related to the ability of people to quickly make decisions in the face of conflicting information—in other words, to resolve conflict.

A flood of observations bear out the association between genes, including those for serotonin, and temperament. Thus, serotonin-related genes could be involved in controlling our emotional disposition at several levels: the production of serotonin within nerve cells, receptors for serotonin (which are coded for by no fewer than 14 genes), serotonin "transporters," which move serotonin from the synaptic cleft back into the nerve cell that had released it, and enzymes that break down the serotonin molecule for later reassembly and reuse. All these levels provide evidence of genetic influence on temperament.

At the level of serotonin production, we must consider the gene coding for the enzyme tryptophan hydroxylase, which is crucial to the synthesis of serotonin. Marc Caron, professor of cell biology at Duke University, found that a mutation in this gene causes an 80 percent drop in the amount of serotonin produced and is associated with major depression in elderly patients. In his study, the mutation showed up much more frequently in the DNA of the depressed patients than in the control group. Even in the latter group, the small number of people carrying the mutation turned out to have come from families with histories of anxiety, alcohol abuse, or mental illness.

A large number of genes for serotonin receptors, too, could be important for temperament. The first studies of these genes, in mice, called for scientists to be able to deduce, from among a given lab population, which individual mice have an anxious temperament. The scientists, Noelia Weisstaub and her collaborators at the Sackler Institute Laboratories of Columbia University, accomplished this by measuring the willingness of each mouse to do things that are risky: entering spaces that are large and open or are well lighted (where predators could see them) or spending time on elevated, suspended beams. According to a 2006 report from the Weisstaub team, mice bearing a genetic mutation that interferes with a specific type of serotonin receptor showed less anxious behavior than the rest of their cohort during these tasks. Here is evidence that this particular receptor plays an important role in the way the brain conjures up the feeling of anxiety. Further evidence comes from the next experimental step: when the scientists carried out an additional molecular manipulation to "undo" the receptor mutation—specifically in the frontal area of the cerebral cortex—the mice once again showed anxiety during the risky tasks. Thinking along similar lines, Silvana

Chiavegatto and her colleagues at the University of São Paulo Medical School, in Brazil, used social isolation of the sort that brings out increased aggression in male mice. This trick, associated with hyperactivity and aggression, led to reduced gene activity for all serotonin receptors that are normally active in the frontal cortex.

As for the next level of the serotonin system, "transporter" is perhaps a misleading term: the protein known by this name doesn't really "transport" serotonin anywhere, but rather it sucks the neurotransmitter out of the synaptic cleft (where it has been stimulating a neuron on the receiving end), and back into the neuron that had originally contained it. Studies in monkeys and humans have involved this gene in the explanation of individual differences in responses to stress. One example comes from a study conducted in female monkeys: Cynthia Bethea, Judy Cameron, and their colleagues at the Oregon National Primate Research Center compared monkeys that were very responsive to stress with those that were relatively unresponsive. They found that in the monkeys' midbrain, in a cluster of serotonin neurons called the dorsal raphe nuclei, the stress-sensitive animals activated fewer copies of the serotonin transporter gene.

Certain variations in the serotonin transporter gene correspond to particular reactions to fearful stimuli occurring in the cerebral cortex. In both nonhuman primates and human infants, one specific version of this gene has been linked with lower abilities to respond to stimuli and higher levels of distress caused by such stimuli. Findings concerning the same gene reinforce our view that genes can make lifelong contributions to temperament, because the above-mentioned mutation affects the behavior of babies as well as of adults. According to a report by John Auerbach and his colleagues in the *Journal of Child Psychology and Psychiatry*, this serotonin transporter gene version in year-old infants is associated with overall negative mood, including expressions of emotional distress when the child's physical activity is gently restrained.

In the United Kingdom, Avshalom Caspi and a large team of scientists in the Institute of Psychiatry at King's College London carried out an epidemiological study that sought to explain why stressful experiences lead to depression in some people but not in others. As they reported in *Science*, the serotonin transporter gene exists in two forms, one short and one long. In response to stress in their lives, people carrying the short version of the serotonin transporter gene showed more depressive symptoms and intentions of suicide than did individuals with only the long version. Since the short version is thought to produce the serotonin transporter less efficiently than the long one, the functional implication of the Caspi study points in the same direction as that of Cynthia Bethea's study in the monkey brain—namely, that normal serotonin transporter levels are needed for an effective response to stressful experiences.

Although Caspi's study still needs to be followed up and replicated, its validity is apparently strengthened by the observations of Turhan Canli, working in the laboratory of Klaus-Peter Lesch at the University of Würzburg in Germany. Canli and Lesch base their study on a growing number of reports that associate the short version of the serotonin transporter gene with high activation of the amygdala, the brain structure we have repeatedly seen to be involved in fear, in certain contexts such as exposure to emotion-laden stimuli or stressful tasks like public speaking. The lower activity of the short gene version appears to enhance brain reactivity to negative stimuli. All these studies reinforce the impression that this particular step in the molecular biology of serotonin signaling—the transporter's removal of serotonin from the synaptic cleft—is central to one major aspect of our temperament: our response to negative emotional events.

Additional genes concerned with removing serotonin from the synaptic cleft take part in the chemical breakdown of the serotonin molecule; chief among these is the gene coding for the enzyme monoamine oxidase A. Caspi and his colleagues have reported that a single modification of this gene causes high aggression. In addition, just as in Bethea's study of monkeys, in which the stress-sensitive group had fewer active copies of this gene at a particular brain site than the stress-resistant group, a study of human volunteers by Andreas Meyer-Lindenberg found that a particular variant of the monoamine oxidase A gene yields only a low level of the enzyme. The deficiency then appears to make the amygdala overresponsive and the regulatory frontal cortex underresponsive, as compared with normal levels. Showing up as they do on structural magnetic resonance imaging scans, these changes fit perfectly with previous observations that the variation in levels of monoamine oxidase A is associated with an increased risk of violent behavior—that is, behavior depending on emotionally labile signals from the amygdala as well as on a lack of inhibition from the frontal cortex.

The brain sites at the receiving end of all these neurotransmitter signals must be numerous indeed, and most remain to be pinpointed. But from all these studies on serotonin, I infer that particular serotonin-producing neurons in the midbrain, both in animals and in people, are transmitting their messages to sites in the ancient forebrain, such as the amygdala, that have to do with the *balance* between negative emotions—stress, threat, or fear—and positive emotions—friendliness, caring, and love.

What about the dispositions toward empathy and affiliative behaviors that we discussed not so long ago? Happily, several genes can be brought back into the discussion here, because those that help to bring about sexual attraction and parental love foster the most primitive, positive sides of human temperament.

A brief recap: the main genes of interest, for our purposes in looking at prosocial behavior, are those involved with the hormones estrogen, prolactin, and oxytocin. From Sonoko Ogawa's work in my laboratory at Rockefeller University, we know that the gene coding for estrogen receptor–alpha is necessary for both female and male sexual behavior. The genes encoding prolactin and the prolactin receptor encourage the protective and nurturing behavior of parenthood, thereby fostering loving behaviors toward other members of the same species as well. And the genes for oxytocin, the main "sociability hormone," and its receptor promote social recognition of the sort that lets an animal, and perhaps a human being, realize that his companion is no threat—that he has had neutral or positive experience with this companion before, and therefore it is safe to be friendly with this guy. All these genes, operating collectively, tend to strengthen the aspects of our temperament that are positive, affiliative, and even hopeful.

A complex puzzle

The study of genetic influences on emotions and temperament is just beginning to take off, and already one thing is clear: the puzzle that we painstakingly assemble will be a complicated one. The old "one gene equals one behavior" formulations were thrown out the window some time ago. Even in my own work with simple behaviors in relatively simple animals like mice, it has been obvious for years that patterns of gene activity influence patterns of behavior. Some of the effects of genes on temperamental dispositions are bound to be complex and indirect. In my book *Brain Arousal and Information Theory,* I list more than 120 genes involved in the fundamental arousal of the central nervous system. Since generalized arousal of the brain is essential to all expressions of emotions, positive or negative, all these genes are more or less implicated in the shaping of temperament.

Frequently, we are talking not just about higher-level cerebral cortical functions but also about lower, visceral contributions to temperament. Furthermore, as we now know, these various genetic alterations interact with one another. Sometimes variations in one gene can have an effect if the variation in the other gene is of a certain sort. Matthew Keller, at UCLA, and Nicholas Martin, at the Queensland Institute of Medical Research in Brisbane, Australia, have collaborated to extend Cloninger's findings on dimensions of temperament; they used twin studies combined with sibling studies to examine how genetic variations interact with one another. Most previous studies had claimed that different genetic influences would simply add to each other—in other words, that one genetic effect would strengthen another; Keller and Martin used the larger databases afforded by including siblings to find evidence for nonadditive genetic effects, which may amplify or diminish one another. Cloninger himself does not analyze his data with a simple formula like, "A change in gene equals a change in behavior." He states clearly that childhood experiences can weigh significantly on the relation between genes and temperament.

Other complexities abound. Sonoko Ogawa, working in my laboratory, has uncovered evidence in mice that the effect of an individual gene on a particular kind of behavior, such as aggression, can depend on at least five distinct factors: exactly when and where that gene is activated in the brain; whether it is activated in a male or a female brain; the age of the animal; the nature of the animal toward which the behavior is directed; and the exact type of aggression measured. And then, as we have seen with serotonin, for every neurotransmitter or other brain chemical that enters the story, we have to consider the genes responsible for its chemical synthesis, its receptor sites, the reentry of the chemical back into the cell once released, and the enzymes that manage its breakdown. Just for dopamine and serotonin, the number of genes involved exceeds two dozen!

Further, John Crabbe, at the Oregon Health and Science University, has emphasized that for some personality attributes, the genetic contributions are not singular or coherent; they can be multiple and disparate. Edward O. Wilson makes a point about difficulties in this field by making the converse argument: with respect to the social behavior of insects, he points to examples in which seemingly unrelated evolutionary phenomena promoting social behavior are actually causally related to each other. In his theoretical argument, the rareness of certain forms of complex societies, the "spectacular ecological success" of these societies, and the forces of natural selection may be causally linked. Even with insect behavior, simpleminded ideas will not carry the day.

Yet for all their complexity, genetic influences alone cannot fully account for a person's temperament or, for that matter, for his or her decision whether to act ethically in a given situation. As I mentioned earlier in this chapter, environmental forces must be taken into consideration to obtain a comprehensive picture of the balancing act that integrates all the various influences and ultimately tips the scales in favor of, or against the Golden Rule.

11

TEMPERAMENT IN THE MAKING

Imagine that you are waiting in line and a stranger tries to cut in front of you. Are you going to let him get away with it, calmly tell him what you think about such behavior, or yell and push him away? Faced with the same circumstances, people respond in an endless variety of ways depending on their temperament and their personality, on who they are. And this personality, in turn, results from elaborate interactions between these environmental influences and the genetic predispositions I discussed in the preceding chapter.

Understanding the interplay between these two types of influences is most pertinent to the study of aggression. In fact, some of the current excitement in my field, the neuroscience of behavior, has to do with figuring out how the interactions between genes and environment work in determining harmfully aggressive behavior. It is a mighty challenge because the complexity of these interactions is enormous.

Thus, when discussing aggression in Chapter 8, I referred to new evidence from Terrie Moffitt and her colleagues at King's College London that maltreatment early in life could "bring out" the effect of a particular gene on violent behavior. Conversely, certain genetic changes render children more likely to respond to early maltreatment by developing conduct disorders. Note that these are statistical trends in human behavior, not statements that a particular gene dominates behavioral choice by a particular individual at a particular time. Moffitt and Michael Rutter of King's College remind us, moreover, that the degree of genetic influence on behavior might itself depend on environmental circumstance. In their words, some behaviors are much more susceptible to social controls than others. In addition, some environments permit the expression of genetically influenced behaviors more than others. As a result, we can expect marked variation in the apparent degree of influence of any given gene over any given human behavior or, for that matter, any given form of psychological abnormality. Moreover, the differences among genetic, inherited, and familial influences must be taken into account.

What determines the impact of environmental forces on our temperament is not only how these forces interact with our genetic makeup, but when this interaction occurs. In early life, all the interactions shaping our temperament are amplified. That is the reason, for example, that in trying to invoke mitigating circumstances for a crime, defense lawyers often cite the harsh treatment or abuse the defendant had suffered in his youth. In fact, there is neuroscientific evidence to prove that two particularly sensitive periods during which environmental factors might produce their most crucial effect are infancy and adolescence. Understanding how our temperaments are shaped at the times that I have called environmental "hot spots" is pertinent to tracing the biological origins of ethical behav-

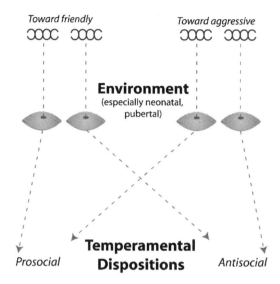

Genes

Toward friendly

Toward aggressive

Environment
(especially neonatal,
pubertal)

**Temperamental
Dispositions**

Prosocial

Antisocial

Figure 11. Temperament. The influences of genes pass through the "lenses" of experience—particularly in the early critical periods of infancy and childhood—to produce the balance between an adult's prosocial and antisocial inclinations.

ior—and perhaps even finding ways to prevent people from violating the Golden Rule.

Sensitivity after birth

In telling you about this fascinating area of research I'd like to start by describing some of the animal experiments proving that certain forms of stress suffered by the newborn can have long-term consequences for their brain and behavior. These, in turn, might affect not only human health but also the "balancing act" that determines whether people are going to act ethically in various situations.

The public's attention was drawn to this area of science during the 1950s by the striking results of Harry Harlow, a professor at the University of Wisconsin, who studied the behavior of rhesus monkeys. At a time when some doctors thought bottle-feeding might be better for human babies, Harlow separated infant monkeys from their mothers and fed them from bottles with formulas intended to improve their health. However, to his surprise, infant mortality among these monkeys actually increased. Further, and more to the point for this chapter, their behavior changed, and these changes lasted beyond infancy.

Monkeys deprived of ongoing contact with their mothers wrapped themselves in their own arms and rocked back and forth, looking pathetic and autistic. Further, in his book *Learning to Love,* Harlow described the problems socially deprived monkeys had with their romantic relations. Rather than approach female monkeys in the spirit and tranquility needed for the female to respond, males who had been deprived of motherly contact—by having been raised in individual wire cages during the ages when infant and peer love normally develops—in Harlow's words, "not only displayed more threatening approaches (to a normal female monkey) but also carried these threats through to damaging physical attacks. These deprived males were totally unable to establish any affectional relations with normal females because they had developed no aggression control through earlier affectional development." Likewise, socially deprived females had trouble establishing good relations with normal males. Instead of engaging in social relations with the male and avoiding physical confrontations, they initiated aggression against males and stopped it only when the larger males fought back. They became depressive, isolated, and socially retarded.

Judy Cameron at the Oregon National Primate Research Center has continued the research into the ways in which the period after birth determines temperament. While her results generally support Harlow's conclusions, they have also revealed the complexity of this subject matter. First, the temperamental responses to external stimuli in monkeys' behaviors are not of a piece. Cameron studied different aspects of the monkeys' anxious and depressive behaviors, comparing responses to novel environments, food, and toys with reactions to threatening social and nonsocial stimuli. She found that the different expressions of anxiety varied depending on the stimuli. Second, different forms of anxious or depressive behaviors are likely to involve a variety of different genes. Cameron has been creating genetic profiles of her monkeys in order to discern which genes are important. Surprisingly, one clue—derived from studies on children described by psychologists as "behaviorally inhibited," that is, fearful, shy, introverted, and afraid to initiate action, especially social interactions—has to do with the growth hormone. The release of this hormone from the pituitary gland can be stimulated by injections of a substance known as the growth hormone releasing hormone. But children and adults who are anxious or depressed have blunted growth hormone responses to this stimulating substance. What about Cameron's monkeys? It turns out that their growth hormone responsiveness to the stimulant also varied depending on their temperaments: young monkeys who had milder reactions to a novel social stimulus had significantly lower responsiveness. These findings provide one possible clue to the mechanisms of behavioral change that might be at play shortly after birth.

Cross-fostering studies with monkeys, furthermore, support our worst suspicion about human infants abused by their mothers: monkeys pass on infant abuse to the next generation; if a female infant is abused by her mother she may grow, in turn, to be a mother who abuses her own babies. Experiments by Dario Maestripieri at the University of Chicago revealed that when a mother monkey who had been abused as an infant goes on to abuse her own offspring, it is the result of her early experience, not her genetic inheritance. What causes her behavior—whether it arises from her social learning or from physiological alterations induced in her body by that ugly early experience—remains to be determined.

Stress hormones

Some of the most striking long-lasting effects of stressful experiences occurring after birth manifest themselves in the way animals handle stress later in life, both in terms of behavior and at the hormonal level. To appreciate the role of hormones in this process, consider first how stress hormones work.

A stressful event—for example, a sudden need to cope with a new situation—triggers a tightly controlled chain of events. First, neurons in the hypothalamus produce the brain chemical called the corticotrophic releasing hormone, or CRH; some of these neurons have long extensions that can deliver this hormone to the pituitary gland. Upon receipt of CRH, certain cells in the pituitary gland release a large protein called the adrenocorticotropic hormone, abbreviated as ACTH, which circulates in the blood outside the brain and arrives at the adrenal glands near the kidneys. Commanded by ACTH, the adrenal glands produce a number of steroidal stress hormones, the better known of which are cortisol in humans and corticosterone in small laboratory animals (and which, just like the CRH, we saw are involved in fear). That's the hormonal stress response in a nutshell: sudden increase in CRH release, causing ACTH release, culminating in adrenal stress hormone release. Since it is important that the hormonal stress response be turned off eventually, the brain has a negative feedback mechanism for keeping it in check: steroid hormones like cortisol and corticosterone send a signal back, both to the brain—to the area called the hippocampus—and to the pituitary gland, to reduce the CRH and ACTH release when the stress subsides.

Because a newborn's contact with the mother constitutes an important factor in the normal development of the hormonal stress response described above, long-term separation from the mother after birth can have long-lasting consequences. A separation of 24 hours, for example, affects the baby animal's nutrition and body temperature and the history of physical contact normally expected from the nursing mother. Paul Plotsky and his colleagues in the Department of Psychiatry at Emory University Medical School have documented impressive increases in the hormonal stress responses in maternally deprived rats. First, CRH release from the hypothalamus is increased in such animals compared to control animals in other litters. It is secreted into its tiny vascular "highway" on the way to the anterior pituitary gland. As a consequence, the release of ACTH is amplified. That causes more corticosterone to pour out of the adrenal gland into the bloodstream. As mentioned above, if this goes on for too long, some types of neural damage might be anticipated. Scientists have found that they can prevent such aberrations in the hormonal stress response, following the absence of the mother, by touching the rat pups in ways that mimic maternal contact. It is the gentle touch and handling of this sort that has helped scientists to start linking experiences in the newborn with the animal's stress response as an adult. Such handling can be assumed to produce a positive effect, as opposed to the bad kind of stress inflicted upon the baby as a result of being separated from the mother.

The first person to investigate the long-term behavioral effects of early handling that mimic certain aspects of the mother's impact on the pups was Seymour Levine, then a professor at Stanford University. One of the effects, produced by brief but systematic gentle handling of rat pups, was to reduce disruptive emotional responses to stressful events later in life, during adulthood. The early handling reduced "emotionality"—in Levine's words, it decreased "diffuse, disorganized, agitated behavior"—as well as expressions of anxiety. For example, rather than being forced to scuttle over to the corner in a lighted field, the animals that had benefited from the gentle handling as pups explored the open spaces of which most rats would be afraid; they also spent more time on the scary open arms of an apparatus called the "elevated plus maze." Moreover, they benefited from an additional prolonged effect of the early stimulation: they quickly learned important responses needed to avoid foot shock. Offering new food for thought is that some of these early-handling effects actually depended on the impact that handling the pups had on their moms, who, as we shall see below, also altered their behavior. In other words, changing the pup changes the mom, whose subsequent interactions with the pup, in turn, permanently change the pup's behavior.

Knowing from the work of Levine and others that gentle handling and stimulation would improve newborn rats' resistance to stress in adulthood, Michael Meaney and his team at the Department of Psychiatry in McGill University School of Medicine in Montreal examined the long-term consequences of these techniques on the pup's adrenal gland responses to stress. Meaney has discovered a mechanism that could account for the improvement: he and colleagues found that the early handling not only decreased the peak levels to which corticosterone can soar in response to stress, but also hastened the hormone's drop to its original, pre-stress levels. This mitigated response may have been due to the changes occurring in the handled animals' brains: the handling could have caused the binding sites for corticosterone in the animals' brains to multiply, thus allowing for a more effective "negative feedback" of stress hormones that turns off the stress response. Meaney's work has shown why the hormonal and emotional changes caused by handling protect the animal over the long term: not only did the stress-protective effects of the early handling persist through all ages tested, but the animals that had been handled after birth were also protected against the damage that tends to occur with age in stress-responsive areas of the brain.

Does the handling produce its effect directly, by affecting the baby's brain, or does it act indirectly, by affecting the mother's behavior? I used to always think the long-term temperamental effects of a newborn's stress depended directly on changes in that baby's brain. That is, brief and careful handling of the baby animal in the laboratory permanently alters the baby's response to stress by virtue of adaptive changes occurring first and foremost in that animal's hypothalamus. The idea was that, just as medical vaccinations could generate acquired immunity against infection, so could "stress inoculation" lead to acquired resistance to stress. However, experiments with laboratory rodents have led to another possibility, the "maternal mediation" hypothesis. Several laboratories have gathered evidence that the increased stress resistance in rats handled after birth really stems from the effect the handling has on the mother: in order to perform the handling, the experimenters had to briefly pick up the pups, and these brief mother-infant separations engendered increased maternal behaviors and grooming—that is, better quality maternal care—throughout the rest of the mother-infant relationship.

After all, in their classic research Vernon Denenberg and Michael Zarrow, then at the University of Illinois, had already demonstrated that greater amounts of *maternal* stimulation received by rat pups during development caused milder stress-induced hormonal responses during adulthood. And their work has been extended by the team of Francesca R. D'Amato at the CNR Institute of Neuroscience in Rome, Italy, whose experiments with mice showed that stressing the mother after delivery by exposing her to the odor of a strange male had an enduring negative effect on adult behavior of the baby mice even without direct manipulation of the pups themselves.

In monkeys—and by extension, in human babies probably too —things are not so simple as the "maternal mediation" hypothesis suggests. Karen Parker and David Lyons, from Stanford University, reported in 2006 in the *Proceedings of the National Academy of Sciences* that compared with rodents, alterations in a monkey mother's care might play less of a role in the baby's subsequent resistance to stress, because intermittent stresses imposed on the infant monkey did not permanently increase maternal care. Moreover, greater amounts of maternal care in these monkeys did not allow the researchers to predict the grown-up offspring's hormonal responses to stress correctly. Instead, intermittent mild stress imposed on the baby monkeys—comprising one-hour separations from the mother once per week—was associated with increased resilience to stress regardless of how the mother was behaving. Thus, in these monkeys, the "stress inoculation" hypothesis held sway over the "maternal mediation" hypothesis, suggesting that stress exposure produced an effect by itself, regardless of the mother's contribution. Systematically conceived and performed studies comparing the relative importance of these two ideas—direct effects on babies versus effects depending on maternal care—remain to be done.

Stressing the mother before delivery seems to affect the pups in a way that is exactly opposite to that of after-birth handling, according to the results of Stefania Maccari at the Université de Bordeaux, France. Thus, after the mothers of experimental rats had been stressed during their last week of pregnancy by being abnormally restrained in a brightly lighted environment, these pups, upon growing up, revealed a slowed-down return of the stress hormone corticosterone to its pre-stress levels. Moreover, these rats were overly anxious, as measured by several tests of animal behavior known to reflect anxiety in rats exposed to things that frighten them, such as open, lighted places, novelty, heights, and so on.

The neural and hormonal responses to stress in the mother animal during her pregnancy and lactation have been studied by Alison Douglas and John Russell at the University of Edinburgh College of Medicine, United Kingdom. As I said in describing the stress cascade, these responses must involve neurons that produce the corticotrophic releasing hormone, or CRH, in the hypothalamus. Normally, the responses of these neurons to stress during pregnancy and lactation are reduced, allowing the mother to stay calm and take good care of her young once they are born. Conversely, as pointed out by Michael Numan and Thomas Insel in *The Neurobiology of Parental Behavior,* "adverse early experiences such as maternal deprivation or separation" can cause an overproduction of CRH in adult females, which in turn increases their fearfulness in a manner that interferes with maternal behavior. These hormonal mechanisms offer yet another route by which early-life stress can affect an adult's temperament.

Most of the experiments with newborn animals have been done with males, but female newborns are also vulnerable to stress. In 2006, Glenda Gillies and her colleagues at Imperial College London and the University of Oxford reported in the journal *Endocrinology* that certain consequences of raised stress hormone levels around the time of birth are different for males and females. In their experiments, female rats—but not males—permanently lost about one-third of their dopamine-producing neurons in the midbrain as a result of brief exposures to increased steroidal stress hormone levels. We do not yet know the exact implications of this neuronal loss for the females' responses to stress later in life compared to those of males, but Gillies's report raises a red flag: given the association of these dopamine neurons with alterations in mood, we must figure out whether untoward experiences soon after birth could contribute to the higher incidence of depressive disorders in women compared with men in adulthood.

We do not yet know exactly how the effects of stress in the newborn translate into enduring, lifelong consequences for health and behavior or what happens at the level of the hormones, brain chemistry, and genes. But I can see parts of the story emerging, and it is clear that temperamental effects deriving from experiences just after birth involve more than fearfulness and resilience.

Beyond fear and anxiety

Prolonged stress can have far-reaching consequences on various body systems—for example, reducing masculine sex drive and performance. But, most relevant for our discussion of ethical behavior, children severely stressed in early childhood are more likely to be violent during adolescence and in young adulthood. W. Thomas Boyce at the Institute of Human Development in the University of California, Berkeley, revealed one way in which children who strongly react to stress can develop temperamental difficulty in later life: these children fall into undesirable company as they tend to be relegated to subordinate positions within social groups characterized by disproportionate rates of medical difficulties, victimization of the children by violence, and disordered social behavior. According to Boyce, marked social subordination not only will make the kid feel lousy but will actually predict later problems with physical and mental health. In Boyce's words, "Youth violence is, at least in part, a developmental progression from peer victimization." In turn, children with conduct disorders are themselves more likely to become parents who raise their own children in a harsh and suboptimal, unforgiving manner.

According to a 2005 article that appeared in *Science* reviewing the major research, the stressful characteristics of a low social rank can have adverse consequences on the cardiovascular, reproductive, immune, and other body systems. The author of the review, Robert Sapolsky, from the Department of Biology at Stanford University, wrote that the research troublingly associated low social rank with increased levels of adrenal stress hormones, which suppress immune responses to health challenges, and with a sluggish recovery from peak stress responses. Equally worrisome are high blood pressure and high cholesterol levels. Sapolsky also lists testicular atrophy and decreased levels of sex hormones in both males and females. And in the brain, he cites impaired generation of new nerve cells, atrophy of nerve cell extensions, and reduced flexibility of neuronal connections.

Several researchers have reported that early social experience can affect brain development, including the development of forebrain neural circuits responsible for regulating the autonomic nervous system and proteins that are known to foster neuronal growth.

However, the exact mechanisms and causal routes of these effects are not understood and are now the subjects of intense investigation. A promising part of the brain to examine is the hippocampus, because it is important for the impact of adrenal stress hormones on the brain's circuitry. Michael Meaney and Moshe Szyf reported in 2006 in the *Proceedings of the National Academy of Sciences* that they had identified more than 900 genes whose activity is affected by early maternal care—a finding that suggests a strong impact of maternal care on the offspring's brain. The scientists compared gene activity in the hippocampus of animals who as pups had been nurtured by moms showing high levels of intense maternal behavior—including licking the pups and assuming nursing postures—with this genetic activity in the hippocampus of animals who had received only low levels of such care. The Meaney team found several differences in gene activity between the two groups, including different activity patterns for genes coding for proteins involved in cellular metabolism, nerve cell energy production, and protein synthesis.

Other researchers have zeroed in on specific nervous system chemicals and the systems in which they act. As reviewed by Charles Marsden from the University of Nottingham Medical School in the United Kingdom, social isolation in developing rats has been associated with changes in the mechanisms involving the neurotransmitter dopamine, which is believed to play a role, among other things, in arousal and aggression; norepinephrine, a hormone involved in arousal, stress, and other body processes; and serotonin, the neurotransmitter that we saw prominently involved in regulating aggressive behaviors. For example, Trevor Robbins at the University of Cambridge, United Kingdom, found that in laboratory animals, rearing in social isolation tends to reduce the release of serotonin in the prefrontal cortex.

From what the studies showed us about maternal love, friendship, and sociability, one would expect to find major differences in the main "sociability hormone," oxytocin, between normally raised

individuals and those who had suffered the stress of maternal neglect shortly after birth. Research, indeed, points in this direction.
A pioneer in the field, Cort Pedersen from the University of North
Carolina, has theorized that parental neglect or abuse of the young
child may cause oxytocin-based human bonding systems to develop
abnormally, causing the affected person to have decreased capacity
for friendships and less commitment to accepted social values. He
cites studies reporting that autistic patients—unable to demonstrate
normal social interest—had lower levels of oxytocin in their blood.

Similarly, extremely low levels of oxytocin were revealed in the
urine of children who were reared in orphanages, according to studies by Wismer Fries that have been analyzed by the expert in the biology of pair bonding, Carol Sue Carter, in the Department of Psychiatry at the University of Illinois at Chicago. A much smaller number
of children reared in regular families had such low levels. Carter is
quick to point out that this difference in oxytocin, as in many other cases, was not necessarily responsible for the differences in social
behavior between the orphaned children and those growing up in
regular families: low oxytocin could either be a cause of such differences, or, conversely, the result of behavioral and other physiological
changes experienced by the orphaned children.

Influence on genes

Can early-childhood experiences have a lifelong effect on the
way genes function in the nervous system? Research suggests that
they can, and that this may be one way in which the sensitivity of
early childhood ends up producing long-lasting effects in behavior.

An early-childhood experience might, for example, permanently turn off certain genes. This idea, put forward by Moshe Szyf, Michael Meaney, and their team at McGill University in Montreal, involves a small chemical "tag" called a methyl group that can modify
one building block of DNA, cytosine. The tag changes the structure
of that part of DNA and effectively turns off, or silences, gene activ-

ity at that site. This gene-silencing "tagging" is believed to be a common occurrence in many types of organisms, and certain evidence supports Szyf's extension of the idea to the early-childhood sensitivity I have been discussing. An important gene for stress hormone signaling to the brain, that encoding the corticosterone receptor, has the kind of structure that favors the attachment of a methyl group, which, in turn, could affect stress responses for most of the life of an animal.

Thus, DNA modification can lead to changes in temperament. Michael Meaney had previously found that early-life stress regulated the activity of the corticosterone gene in the hippocampus, an important brain region for the negative feedback of stress hormones. Further, the Montreal team showed in 2006 that, compared with offspring of high maternal behavior mothers, the offspring of less maternal mothers had more cytosine "tagged" with a methyl group in the DNA region controlling production of an estrogen receptor, and this difference lasted into adulthood. The tagging could mean that the gene would be turned off and the estrogen receptor would no longer be produced. I believe this to be an important finding because estrogen receptor genes appear to have far-reaching effects on behavior: in 1996 Sonoko Ogawa in my lab showed that deleting one such gene could cause females to act like males and be treated like males by their peers. Now there's a marked behavioral change! In coming years I expect that our increased capacity to look at fine molecular changes in small bits of nervous tissue will allow us to pile up evidence that methyl "tagging" of genes turns early-life experiences into a force that has lifelong affects on temperament.

Other ideas, which do not exclude the DNA methyl-"tagging" hypothesis, have been put forward by my colleague at Rockefeller University, David Allis. He has focused on important chemical changes in the proteins that coat DNA—a coating that limits access to the DNA molecule by proteins influencing gene activity. His best-known proposal has to do with a special group of DNA-coating proteins called histones, which can have various molecular tags—methyl groups, phosphate groups, and acetyl groups—added to them as a result of early-life experiences. Relying on his own research and that of other scientists, Allis suggests that the tags create a "barcode" that can alter the activity rate of the genes hidden by the histones for long periods of time. This lasting effect is known as "epigenetic": the genes themselves are not altered, but the way they function is. The finding that the preferred positions of chemical modifications in these DNA-protecting proteins can be predicted—reported in 2006 in *Nature* by Eran Segal at the Weizmann Institute in Israel and Jon Widom at Northwestern University—fits in well with David Allis's ideas.

All the chemical modifications described above can be considered permanent because they can last throughout the lifetime of experimental animals, or for that matter, long enough in humans to influence a person's temperament. Therefore, understanding the sensitivity of the newborn can shed light on some of the mechanisms that later dispose the child and, ultimately, the adult to break or uphold the Golden Rule.

Sensitivity during adolescence

According to Linda Richter of the University of KwaZulu–Natal, South Africa, "Young people in their teens constitute the largest age group in the world, in a special stage recognized across the globe as the link in the life cycle between childhood and adulthood." This age group is characterized by tremendous variability in the timing and degree of physical growth and in the level of social and psychological development.

Similarly to early childhood, the teen years—particularly the years of puberty—are yet another period of heightened sensitivity, when any major influence can have lasting consequences on a person's life. Let us therefore briefly review the hormonal and nervous system changes taking place during this period.

How does puberty work? It is now well established that the primary events responsible for the initiation of mammalian puberty take
place within the brain and not in the pituitary or the sexual glands.
First, the forces driving puberty within the central nervous system
act on the neurons that can make a substance called the luteinizing hormone-releasing hormone, or LHRH, which controls the sex
hormones in both men and women. These neurons, freed from the
inhibitory mechanisms that had until then kept them in check and
stimulated by certain neurotransmitters, start making LHRH. Ser-
gio Ojeda from the Oregon National Primate Research Center is
bringing forth new ideas concerning the involvement of additional
substances and cells that were not linked to the process previously:
Ojeda suggests that growth factors are involved, and that not just
neurons but also the supporting glial cells participate in initiating
puberty. Stimulated by the LHRH, the pituitary gland sends its own
hormones to the testes and ovaries, which start producing testosterone or estrogen. In addition to these rapid changes, which can easily have a disorienting effect on adolescents, increased growth hormone secretion causes the growth spurt that itself can be painful
and confusing.

Pubertal changes in our bodies take place over a much longer
time than most people think. According to Melvin Grumbach, an expert at the medical school of the University of California, San Francisco, puberty can begin as early as eight to nine years of age and in
some individuals not be over until as late as fifteen to sixteen years
of age. Environmental factors such as poor nutrition and stress powerfully influence the time of onset and the rate of hormonal fluc-
tuations of puberty.

Events occurring during puberty will set the stage for the rest of
the child's social life, and social stress during puberty is especially
tough. I am therefore particularly concerned about negative and
enduring effects on a person's adult behavior due to poor treatment at the time of puberty. For boys developing into men, humil-

iation due to low socioeconomic status, anonymity in gargantuan schools, and anomie due to a lack of positive visions for their roles in adult society all predict trouble for the boy's later behaviors.

Long-lasting effects

Neurobiologists are eager to invent models of social stress in pubertal animals so that its consequences for brain function can be studied systematically. Though several such models already exist, only a few major trends can be cited so far, and the emerging picture is rather grim.

Physiological effects of social subordination in puberty include decreased functions of testes and the sexual glands, increased adrenal stress hormone levels, decreased body weight, increased body fat (correlated with heart disease), and decreased lean body mass, according to Randall Sakai of the University of Cincinnati. Brain chemistry is also affected: Sakai describes decreased stress hormone receptors in the hippocampus, which in turn decreases the negative feedback mechanism intended to turn off stress hormone production by the adrenal glands. He also lists several changes in neurotransmitters and an atrophy of the branches of important neurons in the hippocampus, responsible not only for the negative feedback of stress hormones but also for spatial memory.

The effects of stress depend on its duration and on the age at which it is experienced. Russell Romeo, at Rockefeller University, has conducted studies in laboratory animals to distinguish between acute onetime stress, represented by 30 minutes of restraint, and chronic stress: 30 minutes of restraint per day, over 7 days. After acute stress, prepubertal males showed a significantly prolonged hormonal stress response compared with postpubertal males. However, after chronic stress, in prepubertal males the stress hormones peaked at much higher levels and dropped back to their regular, pre-stress levels much faster than in postpubertal animals. In light of these results, which demonstrate prolonged stress consequences of prepubertal stress, it is not surprising that human studies reveal a correlation between stress during adolescence and depression and anxiety during adulthood.

As already mentioned, prolonged stress leads to a prolonged presence of massive levels of the stress hormones cortisol or corticosterone in the blood, which in turn can lead to neuronal degeneration. The harmful changes occur when these hormones bind to glucocorticoid receptors that in turn bind to selected portions of DNA in the nuclei of certain neurons, including important neurons in such parts of the brain as the hippocampus and amygdala. One problem during chronic stress is that it produces a vicious circle: it involves a decrease in the activity of the genes coding for the glucocorticoid receptors, which decreases the negative feedback supposed to turn off the production of the stress hormones, and this leads, in turn, to an even more prolonged stress response—following, for example, separation from the mother.

High, prolonged levels of stress hormones have a deleterious effect on the health of neurons in the hippocampus, as shown years ago by Ana Magarinos in Bruce McEwen's lab at Rockefeller University. This effect could be due to a number of possible mechanisms: altered gene activity in the cell nucleus, a set of rapid stress hormone actions that are initiated at the nerve cell membrane, or the above-mentioned atrophy of the branches of hippocampal neurons. We do not yet know how important these mechanisms are for the damaging effects of stress hormones in the brain, and what is the ultimate significance of the degenerative changes in the hippocampus. Some of these changes are reversible, but how easily, and why? Are some of the apparently degenerative changes actually protective, shielding the brain from stress in ways we do not yet understand? Scientists are currently searching for answers to these questions.

Thus, although we have some clues, we don't really know yet exactly how teenage stress translates into lasting changes of temperament and behavior. From a massive amount of data gathered by Jay

Giedd's team at the National Institute of Mental Health, we clearly see that the adolescent's brain is still undergoing a great deal of the remaining development it needs to do for adulthood and is therefore potentially vulnerable. Apart from the atrophic changes mentioned above, mistakes can occur in the systematic process known as synaptic pruning—a normal and regulated reduction of synaptic contacts at certain sites in the brain at certain stages of development. The teenaged brain is literally a work in progress, and when things are in flux, they can also go wrong rather easily.

The consequences of extreme stress during adolescence are substantial. Kathryn Grant and her colleagues at DePaul University in Chicago have reviewed the evidence that increased stress during childhood and adolescence is associated with increased levels of psychological abnormalities later in life. Two of her conclusions stand out. First, though stress predicts later trouble, particular types of stress risk factors do not predict particular types of psychological outcomes, such as a violent temperament. Second, there is a vicious circle. Just as high stress may lead to psychological abnormalities, displaying deviant behavior may itself cause its own form of stress. For example, serious, persistent hassles and lack of social support during adolescence might lead to depression or inappropriate aggression that in turn would cause additional stress for the young adult.

Changes in temperament during adolescence show up differently in boys and girls. Boys are more likely to begin acting out, getting themselves into trouble and spiraling downward into psychological difficulties in early adulthood. Girls are more likely to get depressed, especially if they are suffering untoward stress during puberty without the benefit of supportive mothers. Elizabeth Young, at the University of Michigan School of Medicine, has hypothesized that the onset of reproductive hormonal changes affecting stress systems at puberty "may sensitize girls to stressful life events, which become more frequent at the transition to puberty and young adulthood."

The stress factors

What societal and other influences of childhood and adolescence contribute to behavior problems later in life? Many psychologists and psychiatrists, including Sven Barnow and Harald Freyberger writing in Mark Mattson's *Neurobiology of Aggression: Understanding and Preventing Violence,* say that children who have been chronically maltreated may be at a higher risk of poorly regulated social behaviors, including an increased risk of inappropriate aggression. Children from broken homes, and those who have suffered physical and sexual abuse, may also be at an increased risk of psychological problems later in life. Moreover, children of women who had been abused are likely to have been as traumatized as if they had been abused themselves.

One of the leaders in showing that stressful environments during puberty can have long-lasting negative consequences for the individual's psychological state is Jacquelynne Eccles in the Institute for Social Research at the University of Michigan. During adolescence, some children drop out of school and their criminal arrest rates rise to the highest level of any age group. These troubling outcomes result, in Eccles's words, "from a mismatch between the needs of developing adolescents and the opportunities afforded them by their social environments."

Eccles and Roberta Simmons have identified the transition from elementary school into middle school (or junior high school) as a risky stage. Junior high schools place a much greater emphasis on discipline imposed by the teacher and less emphasis on providing students with an opportunity to participate in decisions regarding their own learning; there are also fewer personal and positive teacher-student relationships, small group instruction is rare, and individualized instruction may not happen at all. Roberta Simmons feels that "adolescents need a reasonably safe as well as intellectually challenging environment to adapt to these shifts" from elementary to junior high school. If those safe environments are not there, trouble may be just around the corner. Especially for marginal students, Simmons notes, "the early adolescent years mark the beginning of a downward spiral."

In addition to the school-related and other societal factors, physical factors are major contributors to adolescent stress. Beyond the stress of rapid growth mentioned above, all the demands of puberty can, in some individuals, lead to incredible fatigue. In a study in the Netherlands, more than 20 percent of girls and about 6 percent of boys reported significant fatigue. Physicians are concerned that prolonged fatigue may become a more permanent part of those individuals' health and psychological profiles.

Finally, in outlining the long-term effects of chronic stress and other chronic problems on a young person's developing temperament and behavior, I am not ignoring the importance of the immediate situation in which the person is acting. After all, when we talk about "temperamental differences" between two people, we are always describing their different reactions to a provocative situation happening "right now." However, the long-term influences provide the background for the current situation. In defining this background, I identify early adolescent years as a time when the young boy or girl's growth of character and temperament is particularly sensitive to environmental impact.

When intervention is effective

Not all of our measures to prevent teenage troubles need be extremely long or expensive. Geoffrey Cohen and Allison Master, a Colorado/Yale University team, reported in *Science* in 2006 that remarkably brief and inexpensive interventions in the lives of African-American seventh-graders could improve their performance in school. During a fifteen-minute period, the experimenters delivered an in-class writing assignment that was designed to increase students' self-affirmation "by buttressing self-worth" and by "alleviating the stress arising in threatening performance situations."

Students in one experimental group, for example, were asked to write about their most important values, while students in a control group wrote about their least important values. The self-affirmed students' mean grade point averages increased relative to the averages of control students and also relative to those of European-Americans. A key feature of successful interventions is that teenagers must develop more positive perceptions of themselves and their "sense of social connectedness." Failing to provide such guidance means risking a downward spiral—in Cohen's and Master's words, "A negative recursive cycle can occur, where psychological threat and poor performance feed off one another, leading to ever-worsening performance."

The intervention described above is but one of many to emerge from research aimed at preventing aggression and violence. What neuroscience can contribute to their implementation is an understanding of biological mechanisms supporting prosocial and antisocial behaviors—the kind of understanding that the theory I presented in this book can help achieve.

In reviewing all the prosocial and antisocial forces that influence human temperaments, am I surprised at how delicate the balance between these forces could be? Not really. Consider the famous American playwright Sam Shepard and his popular play *True West*. In this play the civil, well-intentioned, highly educated writer Austin is confronted in his mother's empty house by his low-life, violent, criminal brother Lee. Are we just talking about brothers? Not at all. Shepard was quoted in the *Paris Review* as saying that the play reflected the conflicts he felt in his own personality during his visit to his own mother's empty house. If these conflicts between prosocial and antisocial tendencies can happen in the heart and mind of Sam Shepard they can happen in all of us. I hope to have shown in this book that our brains are equipped with all the mechanisms that are needed for the friendly, cooperative tendencies to win.

12

A NEW SCIENCE OF SOCIAL
BEHAVIOR

If you've stayed with me this far, you are now free to admit your initial skepticism toward a theoretical explanation of how humans behave according to the Golden Rule, how aggression is controlled, and how our temperaments affect the balance between altruistic and aggressive forces. It's not really so surprising, considering the subway story with which we began this book. My theory claims that Mr. Autrey's brain must have instantly achieved an identity between his self-image and the image of the victim who fell in front of the subway train. This identification did not occur by some complex, highly intellectual act—it came about by *losing* information, that is, by blurring the distinction between the two images. In addition, Mr. Autrey was demonstrating the kind of prosocial caring feeling that (I hypothesize) normally develops from parental or familial love. In these two ways, the avoidance of harm and the achievement of a helpful act, Mr. Autrey showed his brain to be wired for altruism in the way I hope and believe that all our brains are.

One of the most scientific features of this book—and therefore, to me as a scientist, one of the most exciting—is that it presents clear, testable ideas about how a person puts himself in the place of another. To test these ideas will require brain-imaging experiments in which the activation of specific regions can be compared during tasks that involve ethical decisions. Likewise, electrophysiological studies can look for changes in recordings of the brain's electrical activity during the same tasks. Behavioral approaches can test the idea of reduced discrimination of self from others. Genetic research can examine the predispositions of animals and humans toward affable, generous behavior on one hand, and violent, malicious behavior on the other. With these and other techniques we can examine the fundamental ethical responses I have considered here. We can probably approach more complex questions of social behavior, as well, perhaps even questions we cannot yet conceive of.

The toughest problem I have faced in this book is explaining unethical behavior. It is relatively easy to account for normal, biologically regulated aggression in which animals behave in codified ways to obtain food, attract mates, or defend their territory. The actions of hormones such as testosterone and transmitters such as serotonin take care of a lot. But, in the human case, how are we to understand a man's violence toward women or children? How do we make sense of repeated attempts at genocide? The closest I can come is to hypothesize some kind of collective failure of our natural ability to see others as ourselves, which itself poses major problems for scientists to explain and society to try to relieve.

For now, controversy

I am not alone, of course, in making a solid case for a hypothetical set of brain mechanisms that support ethical behavior. The subject is of great current interest. In February 2007, *New York Times* columnist David Brooks reflected on the ping-pong-like nature of philosophical approaches to ethics. He reminded us that Jean-Jacques Rousseau opined in prerevolutionary France that "everything is good as it leaves the hands of the Author of things"—that is, we are capable of virtuous behavior when in our natural state. Then again, Brooks also quoted the current-day psychologist Steven Pinker as saying, "We strive for dominance and undermine radical egalitarian dreams. We're tribal and divide the world into in-groups and out-groups." Pinker calls this darker view of human nature the "Tragic Vision." I propose that the specific scientific mechanisms underlying ethical behaviors and their transgressions serve to crystallize further thoughts about ethics. They'll be able to bring together anthropological, sociological, economic, and psychological thoughts about these behaviors in a more complete and well-organized way than ever before.

To me, again as a scientist, it is surprising that the treatment of ethical behaviors as a legitimate subject for hard scientific analysis has been as controversial as recent books and articles betray. Currently, in the United States, Daniel C. Dennett, at Tufts University's Department of Philosophy, is leading the charge toward the idea of a scientific explanation of ethical behaviors. In his book *Breaking the Spell,* he calls for "a forthright, scientific, no-holds-barred investigation of religion as one natural phenomenon among many." This is exactly what I have tried to convey in these pages. Dennett undertakes to break down the social and psychological barriers between religious belief and scientific practice; he can easily conceive of the biological value of religion as simply reinforcing prosocial behaviors. Clearly, for him, ethics can be studied as part of natural science, and religious belief can be explained in much the same way as cancer can. The fact that I can offer a coherent hypothesis of the exact brain mechanisms by which people behave in accordance with the Golden Rule offers strong support to Dennett's point of view.

Most unfortunately, Dennett has inspired a counterreaction born of fear and confusion. Leon Wieseltier, the literary editor of the influential political magazine *The New Republic,* is fighting a rearguard battle against biological science. In the *New York Times Book Review,* he distorts the scientific approach by accusing Dennett of "the view that science can explain all human conditions and expressions, mental as well as physical." Wieseltier missed the point that a book like mine will pick a single example of civilized behavior that seems especially amenable to scientific explanation, and then actually explain it. Although in my view Wieseltier is arguing against success, nevertheless Dennett's treatment of religion as a natural phenomenon has put some intellectuals on the defensive. We must hope that continuing education of the public as to the implications of scientific approaches such as the one I have taken here will make supporters of science less nervous.

For the future, challenges

On one hand, human societies have existed for thousands of years, and none of them could have hung together without rules dictating that people treat each other in the way they would like to be treated themselves. On the other hand, right now, according to the World Health Organization, more than 1.5 *million* people are killed each year through violence. I believe these huge social phenomena together reflect, in part, balances and imbalances among competing forces in the human brain. These balances and imbalances operate among various competing economic, social, and political forces in society. The two preceding chapters discussed how, in any individual, balances of temperament between friendly, civil, prosocial forces in the brain and those that produce antisocial, aggressive, and even violent behavior depend on genetic influences, which in turn are powerfully modulated by the environment during early childhood and adolescence. Especially sensitive times for these environmental influences are the periods just after birth and during puberty.

What can a neurobiologist say about *altering* the balance in the brain between forces for civility and opposing forces that lead to violence? Clearly, we all want to increase the strength of brain mechanisms that produce affiliative behaviors and decrease the conditions that would allow mild aggression to explode into violence. First, the disclaimers: there can be no quick fix for aggression, no one-stop shopping for the problems of violence. Nevertheless, there is reason for hope. This book is based on the premise that we are born with mechanisms inherent in our forebrains that dispose us toward behaving in accordance with an ethical universal. Even as Noam Chomsky proposed that we are born with an inherent capacity to learn language, I propose that we have an inherent capacity for fair play. I say we are wired for reciprocity.

Our ability as social beings to tip the balance between prosocial and antisocial influences will be crucial under circumstances in which an individual is at risk to commit violent acts. That is, what neuroscience may achieve in the future is to suggest how to alter brain mechanisms in such a way as to favor affiliative behaviors.

I envision a staged, stepwise approach to the prevention of violence. First, we must admit the gender difference in tendencies toward violence. It is young men, not young women and not an equal ratio of men and women, who commit the larger proportion of aggressive acts of criminality and terrorism, as individuals or en masse. Second, after society has made its best attempts to reduce the devastating effects of the humiliation of extreme socioeconomic differences among families, we need to identify those guys most at risk for untrammeled aggression. This type of estimation will be based on the child's past behavior, his family history, and, perhaps, his genetic profile. What gene variants, a few of which were mentioned in previous chapters, might tell us to pay special attention to a guy's problems? And what are we to do once we have identified those in need of this attention? Certainly behavioral therapy is the way to start. It is less intrusive than most other approaches, and it gives the skilled clinical psychologist the opportunity to understand the nature of each individual's problems. An adolescent boy must be given positive visions of his coming adult role in society; if his family, schools, and/or religion have not fulfilled that vital function, a therapist might still be able to do so. We need rites of passage for teenaged males at risk, buttressed by helpful individual attention.

But also, with our increasing knowledge of how certain sites in the brain control fear and aggression, we are heading for a better neurochemistry—and thus better behavior-altering drugs—combined with a better understanding of genetic variations that are associated with increased risk of violence or criminality. Thus, we're making progress toward being able to custom-design a unique medical regimen for any individual patient on the basis of his genetic profile—an approach that might head off tendencies toward the worst behaviors in the most vulnerable guys.

As of right now, which genetic systems would I single out for special attention in such an approach? Three systems in particular would, I think, be worth a good, hard look. First, oxytocin. The genes coding for this brain chemical and its receptor powerfully promote friendly behaviors; let's figure out how to ramp them up. Second, working in the opposite direction, we must reduce the effects of male sex hormones in some individuals, because the cascades of signals set off by testosterone and the neurochemical changes that follow from testosterone-related gene activity are certainly associated with peaks of violent behavior during the lifetime of some males. Third, all the fear and aggression, whether carried out by individuals or by groups, is fueled by the generalized arousal systems in the brain. If a guy is simply set on a hair trigger, we must use the tools of modern neurochemistry to quiet him down, carefully and measurably.

Here is a persuasive reason for hope and optimism: from a technical point of view, considering forebrain mechanisms that offer friendliness versus those that facilitate aggression, I believe that relatively small changes, well chosen, will *tip the balance* toward prosocial behavior in the individual young man.

Along with these specific neuroscientific suggestions, I have three more caveats. First, neuroscientists are not looking for a perfect society, as has sometimes been charged. They simply want to head off some of the worst problems that can be tied to the behavioral outputs of individuals' brains. Second, the balance between prosocial and antisocial mechanisms in the brain is sensitive and fragile; most of all, it is dynamic. Thus, attempts at helpful intervention must be monitored continuously. Finally, what my neuroscientific theory cannot address are all the social, economic, and political events that magnify ugly tendencies toward aggression by the individual.

It seems amazing that modern neurobiology now approaches an explanation of behavior in accordance with an ethical principle that has become universal in recent social history—the principle of treating others as we would wish ourselves to be treated. Of course, even if my scientific explanation provides an understanding of brain mechanisms underlying an ethical behavior, it does nothing more than supplement other intellectual approaches to ethical behaviors. Neuroscience is by no means a replacement but only a new partner for other intellectual approaches to the discipline of ethics. But the more we know about brain mechanisms for empathy and ethical behaviors, the more effectively we can deal with situations in which we are saddened by their absence.

I am optimistic. Despite the military actions and political troubles we read about in our newspapers every day, I think the inherent disposition in the human brain toward civil, ethical behavior will win out. At a Live 8 conference in Philadelphia on July 4, 2006, *New York Times* music critic Jon Pareles wrote that while the event commemorated the American Declaration of Independence—the performers called for a "declaration of *inter*dependence." What could be more true on today's shrinking planet, with our international economy and our cultural worlds without borders? President Michelle Bachelet of Chile, whose father died under politically motivated torture, showed the way when she said recently, "Because I was the victim of hatred, I have dedicated my life to reverse that hatred and turn it into understanding, tolerance and—why not say it—love."

REFERENCES

Chapter 2, The Golden Rule (pages 7 to 20)

Axelrod, R. *The Evolution of Cooperation.* New York: Basic Books, 1984.
———. *The Complexity of Cooperation: Agent-Based Models of Competition and Collaboration.* Princeton: Princeton University Press, 1997.
Axelrod, R., and W. D. Hamilton. "The Evolution of Cooperation." *Science* 211 (1981): 1390–1396.
Das, B., comp. *The Essential Unity of All Religions.* 3rd ed. Benares: Ananda Publishing House, 1947.
Doebeli, M., et al. "The Evolutionary Origin of Cooperators and Defectors." *Science* 306 (2004): 859–862.
Gazzaniga, M. S. *The Ethical Brain.* New York: Dana Press, 2005.
Greene, J. D., et al. "An fMRI Investigation of Emotional Engagement in Moral Judgment." *Science* 293 (2001): 2105–2108.
Hauser, M. D., et al. "Give unto Others: Genetically Unrelated Cotton-Top Tamarin Monkeys Preferentially Give Food to Those Who Altruistically Give Food Back." *Proceedings of the Royal Society of London,* ser. B, 270 (2003): 2363–2370.
Hinde, R. A. *Why Gods Persist: A Scientific Approach to Religion.* London: Routledge, 1999.
———. *Why Good Is Good: The Sources of Morality.* London: Routledge, 2002.
King-Casas, B., et al. "Getting to Know You: Reputation and Trust in a Two-Person Exchange." *Science* 308 (2005): 78–83.
Kummer, H. "Analogs of Morality among Nonhuman Primates." In *Morality as a Biological Phenomenon: The Pre-suppositions of Sociobiological Research,* edited by G. S. Stent, pp. 31–49. Berkeley: University of California Press, 1978.
Pfaff, D. W. *Drive: Neurobiological and Molecular Mechanisms of Sexual Motivation.* Cambridge: MIT Press, 1999.
Rottschaefer, W. A. "Naturalizing Ethics: The Biology and Psychology of Moral Agency." *Zygon* 35 (2000): 253–286.
Singer, T., et al. "Empathy for Pain Involves the Affective but Not Sensory

Components of Pain." *Science* 303 (2004): 1157–1162.

Wilson, E. O. *On Human Nature.* Cambridge: Harvard University Press, 1978.

Chapter 3, Being Afraid (pages 21 to 39)

Buss, K. A., et al. "Context-Specific Freezing and Associated Physiological Reactivity as a Dysregulated Fear Response." *Developmental Psychology* 40 (2004): 583–594.

Dalton, K. M., et al. "Neural-Cardiac Coupling in Threat-Evoked Anxiety." *Journal of Cognitive Neuroscience* 17 (2005): 969–980.

Davis, M. "Searching for a Drug to Extinguish Fear." *Cerebrum* 7 (2005): 47–58.

de Gelder, B., et al. "Unconscious Fear Influences Emotional Awareness of Faces and Voices." *PNAS* 102 (2005): 18682–18687.

Duvarci, S., et al. "Activation of Extracellular Signal-Regulated Kinase-Mictogen-Activated Protein Kinase Cascade in the Amygdala Is Required for Memory Reconsolidation of Auditory Fear Conditioning." *European Journal of Neuroscience* 21 (2005): 283–289.

Kalin, N. H., et al. "The Role of the Central Nucleus of the Amygdala in Mediating Fear and Anxiety in the Primate." *Journal of Neuroscience* 24 (2004): 5506–5515.

LeDoux, J. E. "Emotion Circuits in the Brain." *Annual Review of Neuroscience* 23 (2000): 155–184.

Malin, E. L., and J. L. McGaugh. "Differential Involvement of the Hippocampus, Anterior Cingulate Cortex, and Basolateral Amygdala in Memory for Context and Footshock." *PNAS* 103 (2006): 1959–1963.

Pfaff, D. *Brain Arousal and Information Theory: Neural and Genetic Mechanisms.* Cambridge: Harvard University Press, 2006.

Phelps, E. A., and J. E. LeDoux. "Contributions of the Amygdala to Emotion Processing: From Animal Models to Human Behavior." *Neuron* 48 (2005): 175–187.

Ressler, K. J., et al. "Cognitive Enhances as Adjuncts to Psychotherapy: Use of D-Cycloserine in Phobic Individuals to Facilitate Extinction of Fear." *Archives of General Psychiatry* 61 (2004): 1136–1144.

Urry, H. L., et al. "Making a Life Worth Living: Neural Correlates of Well-Being." *Psychological Science* 15 (2004): 367–372.

Chapter 4, Memory of Fear (pages 40 to 60)

Fanselow, M. S., and G. D. Gale. "The Amygdala, Fear, and Memory." In *The Amygdala in Brain Function: Basic and Clinical Approaches,* edited by P. Shinnick-Gallagher et al., 125–134. Annals of the New York Academy of Sciences 985. New York: New York Academy of Sciences, 2003.

Gale, G. D., et al. "Role of the Basolateral Amygdala in the Storage of Fear Memories across the Adult Lifetime of Rats." *Journal of Neuroscience* 24 (2004): 3810–3815.

McEwen, B. S., with E. N. Lasley. *The End of Stress as We Know It.* Washington, DC: Dana Press, 2004.

McIntyre, C. K., et al. "Memory-Influencing Intra-basolateral Amygdala Drug

Infusions Modulate Expression of Arc Protein in the Hippocampus."
PNAS 102 (2005): 10718–10723.

Paton, J. J., et al. "The Primate Amygdala Represents the Positive and Negative Value of Visual Stimuli during Learning." *Nature* 439 (2006): 865–870.

Pessoa, L., and S. Padmala. "Quantitative Prediction of Perceptual Decisions during Near-Threshold Fear Detection." *PNAS* 102 (2005): 5612–5617.

Pfaff, D. *Brain Arousal and Information Theory: Neural and Genetic Mechanisms.* Cambridge: Harvard University Press, 2006.

Rodrigues, S. M., et al. "Molecular Mechanisms Underlying Emotional Learning and Memory in the Lateral Amygdala." *Neuron* 44 (2004): 75–91.

Roozendaal, B., E. L. Hahn, et al. "Glucocorticoid Effects on Memory Retrieval Require Concurrent Noradrenergic Activity in the Hippocampus and Basolateral Amygdala." *Journal of Neuroscience* 24 (2004): 8161–8169.

Roozendaal, B., S. Okuda, et al. "Glucocorticoids Interact with Emotion-Induced Noradrenergic Activation in Influencing Different Memory Functions." *Neuroscience* 138 (2006): 901–910.

Schafe, G. E., et al. "Tracking the Fear Engram: The Lateral Amygdala Is an Essential Locus of Fear Memory Storage." *Journal of Neuroscience* 25 (2005): 10010–10014.

Shumyatsky, G. P., et al. "*Stathmin,* a Gene Enriched in the Amygdala, Controls Both Learned and Innate Fear." *Cell* 123 (2005): 697–709.

Watts, A. G. "Glucocorticoid Regulation of Peptide Genes in Neuroendocrine CRH Neurons: A Complexity beyond Negative Feedback." *Frontiers in Neuroendocrinology* 26 (2005): 109–130.

Chapter 5, Losing Oneself (pages 61 to 79)

Bandyopadhyay, S., et al. "Endogenous Acetylcholine Enhances Synchronized Interneuron Activity in Rat Neocortex." *Journal of Neurophysiology* 95 (2006): 1908–1916.

Carr, L., et al. "Neural Mechanisms of Empathy in Humans: A Relay from Neural Systems for Imitation to Limbic Areas." *PNAS* 100 (2003): 5497–5502.

Desimone, R., et al. "Stimulus-Selective Properties of Inferior Temporal Neurons in the Macaque." *Journal of Neuroscience* 4 (1984): 2051–2062.

de Waal, F. B. M., et al. "The Monkey in the Mirror: Hardly a Stranger." *PNAS* 102 (2005): 11140–11147.

Gross, C. G., et al. "Visual Properties of Neurons in Inferotemporal Cortex of the Macaque." *Journal of Neurophysiology* 35 (1972): 96–111.

Held, R., and S. J. Freedman. "Plasticity in Human Sensorimotor Control." *Science* 142 (1963): 455–462.

Kanwisher, N., et al. "The Fusiform Face Area: A Module in Human Extrastriate Cortex Specialized for Face Perception." *Journal of Neuroscience* 17 (1997): 4302–4311.

Knapska, E., et al. "Between-Subject Transfer of Emotional Information Evokes Specific Pattern of Amygdala Activation." *PNAS* 103 (2006): 3858–3862.

Lau, H. C., et al. "Attention to Intention." *Science* 303 (2004): 1208–1210.

McKinstry, J. L., et al. "A Cerebellar Model for Predictive Motor Control Tested in a Brain-Based Device." *PNAS* 103 (2006): 3387–3392.

PAIN 125 (2006) 5–9. Topical Review by P. L. Jackson, P. Rainville, J. Decety. To what extent do we share the pain of others? Insight from the neural basis of pain empathy.

Poulet, J. F. A., and B. Hedwig. "The Cellular Basis of a Corollary Discharge." *Science* 311 (2006): 518–522.

Quiroga, R. Q., et al. "Movement Intention Is Better Predicted than Attention in the Posterior Parietal Cortex." *Journal of Neuroscience* 26 (2006): 3615–3620.

Chapter 6, Sex and Parental Love (pages 80 to 98)

Fisher, H. *Why We Love: The Nature and Chemistry of Romantic Love.* New York: Henry Holt, 2004.

Gonzalez-Mariscal, G., and P. Poindron. "Parental Care in Mammals: Immediate Internal and Sensory Factors of Control." In *Hormones, Brain, and Behavior,* 5 vols., edited by D. W. Pfaff et al., 1:215–298. San Diego: Academic Press (Elsevier), 2002.

Numan, M., and T. R. Insel. *The Neurobiology of Parental Behavior.* Heidelberg: Springer Verlag, 2003.

Pfaff, D. W. *Drive: Neurobiological and Molecular Mechanisms of Sexual Motivation.* Cambridge: MIT Press, 1999.

Roughgarden, J., et al. "Reproductive Social Behavior: Cooperative Games to Replace Sexual Selection." *Science* 311 (2006): 965–969.

Winslow, J. T., et al. "A Role for Central Vasopressin in Pair Bonding in Monogamous Prairie Voles." *Nature* 365 (1993): 545–548.

Young, L. J., et al. "Cellular Mechanisms of Attachment." *Hormones and Behavior* 40 (2001): 133–138.

Chapter 7, Sociability (pages 99 to 120)

Carter, C. S., and E. B. Keverne. "The Neurobiology of Social Affiliation and Pair Bonding." In *Hormones, Brain, and Behavior,* 5 vols., edited by D. W. Pfaff et al., 1:299–337. San Diego: Academic Press (Elsevier), 2002.

Choleris, E., et al. "An Estrogen-Dependent Four-Gene Micronet Regulating Social Recognition: A Study with Oxytocin and Estrogen Receptor–alpha and –beta Knockout Mice." *PNAS* 100 (2003): 6192–6197.

Curley, J. P., and E. B. Keverne. "Genes, Brains and Mammalian Social Bonds." *Trends in Ecology and Evolution* 20 (2005): 561–567.

Goodson, J. L. "The Vertebrate Social Behavior Network: Evolutionary Themes and Variations." *Hormones and Behavior* 48 (2005): 11–22.

Jansen, V. A. A., and M. van Baalen. "Altruism through Beard Chromodynamics." *Nature* 440 (2006): 663–666.

Kirsch, P., et al. "Oxytocin Modulates Neural Circuitry for Social Cognition and Fear in Humans." *Journal of Neuroscience* 25 (2005): 11489–11493.

Numan, M., and T. R. Insel. *The Neurobiology of Parental Behavior.* Heidelberg: Springer Verlag, 2003.

Ogawa, S., et al. "Roles of Estrogen Receptor–alpha Gene Expression in Re-production-Related Behaviors in Female Mice." *Endocrinology* 139 (1998): 5070–5081.

Singer, T., S. J. Kiebel, et al. "Brain Responses to the Acquired Moral Status of Faces." *Neuron* 41 (2004): 653–662.

Singer, T., B. Seymour, et al. "Empathic Neural Responses Are Modulated by the Perceived Fairness of Others." *Nature* 439 (2006): 466–469.

Wilson, J. Q. *The Moral Sense.* New York: Macmillan (Free Press), 1993.

Zak, P. J., et al. "Oxytocin Is Associated with Human Trustworthiness." *Hormones and Behavior* 48 (2005): 522–527.

Chapter 8, The Urge to Harm (pages 121 to 144)

Albers, H. E., et al. "Hormonal Basis of Social Conflict and Communication." In *Hormones, Brain, and Behavior,* 5 vols., edited by D. W. Pfaff et al., 1:393–433. San Diego: Academic Press (Elsevier), 2002.

Berkowitz, L. "Affect, Aggression, and Antisocial Behavior." In *Handbook of Affective Sciences,* edited by R. J. Davidson et al., 804–823. Oxford: Oxford University Press, 2003.

Bock, G., and J. Goode, eds. *Molecular Mechanisms Influencing Aggressive Behaviours.* Novartis Foundation Symposium 268. Chichester, UK: Wiley & Sons, 2005.

Caspi, A., et al. "Role of Genotype in the Cycle of Violence in Maltreated Children." *Science* 297 (2002): 851–854.

De Vries, G. J., and G. C. Panzica. "Sexual Differentiation of Central Vasopressin and Vasotocin Systems in Vertebrates: Different Mechanisms, Similar Endpoints." *Neuroscience* 138 (2006): 947–955.

Ehrenreich, B. *Blood Rites: Origins and History of the Passions of War.* New York: Henry Holt, 1997.

Hrabovszky, E., et al. "Neurochemical Characterization of Hypothalamic Neurons Involved in Attack Behavior: Glutamatergic Dominance and Co-expression of Thyrotopin-Releasing Hormone in a Subset of Glutamatergic Neurons." *Neuroscience* 133 (2005): 657–666.

Kagan, D. *On the Origins of War and the Preservation of Peace.* New York: Doubleday, 1995.

Kelly, R. C. "The Evolution of Lethal Intergroup Violence." *PNAS* 102 (2005): 15294–15298.

———. *Warless Societies and the Origin of War.* Ann Arbor: University of Michigan Press, 2000.

Nelson, R. J., ed. *Biology of Aggression.* New York: Oxford University Press, 2006.

Pfaff, D. *Brain Arousal and Information Theory: Neural and Genetic Mechanisms.* Cambridge: Harvard University Press, 2006.

Simon, N. G. "Hormonal Processes in the Development and Expression of Aggressive Behavior." In *Hormones, Brain, and Behavior,* 5 vols., edited by D. W. Pfaff et al., 1:339–392. San Diego: Academic Press (Elsevier), 2002.

Tuchman, B. W. *The March of Folly: From Troy to Vietnam.* New York: Ballantine Books, 1985.

Vasudevan, N., et al. "Early Membrane Estrogenic Effects Required for Full Expression of Slower Genomic Actions in a Nerve Cell Line." *PNAS* 98 (2001): 12267–12271.

———. "Integration of Steroid Hormone Initiated Membrane Action to Genomic Function in the Brain." *Steroids* 70 (2005): 388–396.

Chapter 9, Murder and Other Mayhem (pages 145 to 160)

Côté, S., et al. "The Development of Physical Aggression from Toddlerhood to Pre-adolescence: A Nation Wide Longitudinal Study of Canadian Children." *Journal of Abnormal Child Psychology* 34 (2006): 71–85.

Crowell, N. A., and A. W. Burgess. *Understanding Violence Against Women.* Washington, DC: National Academy Press, 1996.

Davis, M. "Neural Systems Involved in Fear-Potentiated Startle." In *Modulation of Defined Vertebrate Neural Circuits,* ed. M. Davis et al., 165–183. Annals of the New York Academy of Sciences 563. New York: New York Academy of Sciences, 1989.

Devine, J., et al., eds. *Youth Violence: Scientific Approaches to Prevention.* Annals of the New York Academy of Sciences 1036. New York: New York Academy of Sciences, 2004.

Echeburúa, E., et al. "Psychopathological Profile of Men Convicted of Gender Violence: A Study in the Prisons of Spain." *Journal of Interpersonal Violence* 18 (2003): 798–812.

Gilligan, J. *Preventing Violence.* London: Thames & Hudson, 2001.

Hagedorn, J., ed. *Gangs in the Global City: Alternatives to Traditional Criminology.* Champaign: University of Illinois Press, 2007.

Henderson, L. P., et al. "Anabolic Androgenic Steroids and Forebrain GABAergic Transmission." *Neuroscience* 138 (2006): 793–799.

Levenston, G. K., et al. "The Psychopath as Observer: Emotion and Attention in Picture Processing." *Journal of Abnormal Psychology* 109 (2000): 373–385.

McGinnis, M. Y. "Anabolic Androgenic Steroids and Aggression: Studies Using Animal Models." In *Youth Violence: Scientific Approaches to Prevention,* ed. J. Devine et al., 399–415. Annals of the New York Academy of Sciences 1036. New York: New York Academy of Sciences, 2004.

Ogawa, S., V. Eng., et al. "Roles of Estrogen Receptor–alpha Gene Expression in Reproduction-Related Behaviors in Female Mice." *Endocrinology* 139 (1998): 5070–5081.

Ogawa, S., T. F. Washburn, et al. "Modifications of Testosterone-Dependent Behaviors by Estrogen Receptor–alpha Gene Disruption in Male Mice." *Endocrinology* 139 (1998): 5058–5069.

Patrick, C. J., et al. "Emotion in the Criminal Psychopath: Startle Reflex Modulation." *Journal of Abnormal Psychology* 102 (1993): 82–92.

Pfaff, D. *Brain Arousal and Information Theory: Neural and Genetic Mechanisms.* Cambridge: Harvard University Press, 2006.

Pope, H. G., Jr., and D. L. Katz. "Homicide and Near-Homicide by Anabolic Steroid Users." *Journal of Clinical Psychiatry* 51 (1990): 28–31.

Reiss, A. J., Jr., and J. A. Roth, eds. *Understanding and Preventing Violence.* A National Research Council report. Washington, DC: National Academy Press, 1993.

Reiss, A. J., Jr., et al., eds. *Understanding and Preventing Violence.* Vol. 2. Washington, DC: National Academy Press, 1994.

Su, T-P, et al. "Neuropsychiatric Effects of Anabolic Steroids in Male Normal Volunteers." *JAMA* 269 (1993): 2760–2764.

Chapter 10, Balancing Act (pages 161 to 178)

Bethea, C. L., et al. "Serotonin-Related Gene Expression in Female Monkeys with Individual Sensitivity to Stress." *Neuroscience* 132 (2005): 151–166.

Canli, T., et al. "Beyond Affect: A Role for Genetic Variation of the Serotonin Transporter in Neural Activation during a Cognitive Attention Task." *PNAS* 102 (2005): 12224–12229.

Cloninger, C. R. *Feeling Good: The Science of Well-Being.* New York: Oxford University Press, 2004.

de Laszlo, V. S., ed. *The Basic Writings of C. G. Jung.* New York: Modern Library, 1993.

Hall, C. S., et al. *Theories of Personality.* 4th ed. New York: Wiley & Sons, 1998.

Kagan, J. *Galen's Prophecy: Temperament in Human Nature.* New York: Basic Books, 1994.

———. *Surprise, Uncertainty, and Mental Structures.* Cambridge: Harvard University Press, 2002.

Kagan, J., and N. Snidman. *The Long Shadow of Temperament.* Cambridge: Harvard University Press, 2004.

Keller, M. C., et al. "Widespread Evidence for Non-additive Genetic Variation in Cloninger's and Eysenck's Personality Dimensions Using a Twin plus Sibling Design." *Behavior Genetics* 35 (2005): 707–721.

Kreek, M. J., et al. "Genetic Influences on Impulsivity, Risk Taking, Stress Responsivity and Vulnerability to Drug Abuse and Addiction." *Nature Neuroscience* 8 (2005): 1450–1457.

Lumsden, C. J., and E. O. Wilson. *Genes, Mind, and Culture: The Coevolutionary Process.* Cambridge: Harvard University Press, 1981.

Meyer-Lindenberg, A., et al. "Neural Mechanisms of Genetic Risk for Impulsivity and Violence in Humans." *PNAS* 103 (2006): 6269–6274.

Ogawa, S., et al. "Genetic Influences on Aggressive Behaviors and Arousability in Animals." In *Youth Violence: Scientific Approaches to Prevention,* edited by J. Devine et al., 257–266. Annals of the New York Academy of Sciences 1036. New York: New York Academy of Sciences, 2004.

Raine, A., et al. "Neurocognitive Impairments in Boys on the Life-Course Persistent Antisocial Path." *Journal of Abnormal Psychology* 114 (2005): 38–49.

Rothbart, M. K., and J. E. Bates. "Temperament." In *Handbook of Child Psychology,* 6th ed., edited by W. Damon and R. M. Lerner, vol. 3, *Social, Emotional, and Personality Development,* edited by N. Eisenberg, 99–166. New York: Wiley & Sons, 2006.

Schwartz, C. E., et al. "Inhibited and Uninhibited Infants 'Grown Up': Adult Amygdala Responses to Novelty." *Science* 300 (2003): 1952–1953.

Serretti, A., et al. "Temperament and Character in Mood Disorders: Influence of DRD4, SERTPR, TPH and MAO-A Polymorphisms." *Neuropsychobiology* 53 (2006): 9–16.

Weisstaub, N. V., et al. "Cortical 5-HT$_{2A}$ Receptor Signaling Modulates Anxiety-Like Behaviors in Mice." *Science* 313 (2006): 536–540.

Wilson, E. O., and B. Hölldobler. "Eusociality: Origin and Consequences." *PNAS* 102 (2005): 13367–13371.

Chapter 11, Temperament in the Making (pages 179 to 201)

Champagne, F. A., et al. "Maternal Care Associated with Methylation of the Estrogen Receptor–alpha1b Promoter and Estrogen Receptor–alpha Expression in the Medial Preoptic Area of Female Offspring." *Endocrinology* 147 (2006): 2909–2915.

Dallman, M. F. "Fast Glucocorticoid Actions on Brain: Back to the Future." *Frontiers in Neuroendocrinology* 26 (2005): 103–108.

Eccles, J. S., et al. "Development during Adolescence: The Impact of Stage-Environment Fit on Young Adolescents' Experiences in Schools and in Families." *American Psychologist* 48 (1993): 90–101.

Hake, S. B., and C. D. Allis. "Histone H3 Variants and Their Potential Role in Indexing Mammalian Genomes: The 'H3 Barcode Hypothesis.'" *PNAS* 103 (2006): 6428–6435.

Levine, S. "Enduring Effects of Early Experience on Adult Behavior." In *Hormones, Brain, and Behavior*, 5 vols., edited by D. W. Pfaff et al., 4:535–542. San Diego: Academic Press (Elsevier), 2002.

Mattson, M. P., ed. *Neurobiology of Aggression: Understanding and Preventing Violence*. Totowa, NJ: Humana Press, 2003.

McEwen, B. S. "Stress and Hippocampal Plasticity." *Annual Review of Neuroscience* 22 (1999): 105–122.

Ojeda, S. R., and E. Terasawa. "Neuroendocrine Regulation of Puberty." In *Hormones, Brain, and Behavior*, 5 vols., edited by D. W. Pfaff et al., 4:589–659. San Diego: Academic Press (Elsevier), 2002.

Romeo, R. D., et al. "Stress History and Pubertal Development Interact to Shape Hypothalamic-Pituitary-Adrenal Axis Plasticity." *Endocrinology* 147 (2006): 1664–1674.

Styne, D. M., and M. Grumbach. "Puberty in Boys and Girls." In *Hormones, Brain, and Behavior*, 5 vols., edited by D. W. Pfaff et al., 4:661–716. San Diego: Academic Press (Elsevier), 2002.

Szyf, M., et al. "Maternal Programming of Steroid Receptor Expression and Phenotype through DNA Methylation in the Rat." *Frontiers in Neuroendocrinology* 26 (2005): 139–162.

Walker, C.-D., et al. "Glucocorticoids, Stress and Development." In *Hormones, Brain, and Behavior*, 5 vols., edited by D. W. Pfaff et al., 4:487–534. San Diego: Academic Press (Elsevier), 2002.

Chapter 12, A New Science of Social Behavior (pages 202 to 209)

Dennett, D. C. *Breaking the Spell: Religion as a Natural Phenomenon*. New York: Viking, 2006.

INDEX

Page numbers in *italics* indicate illustrations.

OTHER DANA PRESS BOOKS & PERIODICALS

www.dana.org

Books for General Readers

BRAIN & MIND

BEST OF THE BRAIN FROM SCIENTIFIC AMERICAN:
Mind, Matter, and Tomorrow's Brain

Floyd E. Bloom, M.D., Editor

Top neuroscientist Floyd E. Bloom has selected the most fascinating brain-related articles from *Scientific American* and *Scientific American Mind* since 1999 in this collection. Divided into three sections—Mind, Matter, and Tomorrow's Brain—this compilation offers the latest information from the front lines of brain research. 30 full-color illustrations.

Cloth, 261 pp. 1-932594-22-1 • $25.00
ISBN-13: 978-1-932594-22-5

CEREBRUM 2007: Emerging Ideas in Brain Science

Cynthia A. Read, Editor

Foreword by Bruce McEwen, Ph.D.

Prominent scientists and other thinkers explain, applaud, and protest new ideas arising from discoveries about the brain in this first yearly anthology from *Cerebrum's* Web journal for inquisitive general readers. 10 black-and-white illustrations.

Paper 243 pp. 978-1-932594-24-9 • $14.95

MIND WARS: Brain Research and National Defense

Jonathan Moreno, Ph.D.

A leading ethicist examines national security agencies' work on defense applications of brain science, and the ethical issues to consider.

Cloth 210 pp. 1-932594-16-7 • $23.95
ISBN-13: 978-1-932594-16-4

THE DANA GUIDE TO BRAIN HEALTH: A Practical Family Reference
from Medical Experts (with CD-ROM)

Floyd E. Bloom, M.D., M. Flint Beal, M.D., and David J. Kupfer, M.D., Editors

Foreword by William Safire

The only complete, authoritative family-friendly guide to the brain's development, health, and disorders. *The Dana Guide to Brain Health* offers ready reference to our latest understanding of brain diseases as well as information to help you participate in your family's care. 16 full-color illustrations and more than 200 black-and-white drawings.

Paper (with CD-ROM) 733 pp. 1-932594-10-8 • $25.00
ISBN-13: 978-1-932594-10-2

THE CREATING BRAIN: *The Neuroscience of Genius*

Nancy C. Andreasen, M.D., Ph.D.

A noted psychiatrist and bestselling author explores how the brain achieves creative breakthroughs, including questions such as how creative people are different and the difference between genius and intelligence. She also describes how to develop our creative capacity. 33 illustrations/photos.

Cloth 197 pp. 1-932594-07-8 • $23.95
ISBN-13: 978-1-932594-07-2

THE ETHICAL BRAIN

Michael S. Gazzaniga, Ph.D.

Explores how the lessons of neuroscience help resolve today's ethical dilemmas, ranging from when life begins to free will and criminal responsibility. The author, a pioneer in cognitive neuroscience, is a member of the President's Council on Bioethics.

Cloth 201 pp. 1-932594-01-9 • $25.00

A GOOD START IN LIFE: Understanding Your Child's Brain and Behavior
from Birth to Age 6

Norbert Herschkowitz, M.D., and Elinore Chapman Herschkowitz

The authors show how brain development shapes a child's personality and behavior, discussing appropriate rule-setting, the child's moral sense, temperament, language, playing, aggression, impulse control, and empathy. 13 illustrations.

Cloth 283 pp. 0-309-07639-0 • $22.95
Paper (Updated with new material) 312 pp. 0-9723830-5-0 • $13.95

BACK FROM THE BRINK: How Crises Spur Doctors to New Discoveries about the Brain

Edward J. Sylvester

In two academic medical centers, Columbia's New York Presbyterian and Johns Hopkins Medical Institutions, a new breed of doctor, the neurointensivist, saves patients with life-threatening brain injuries. 16 illustrations/photos.

Cloth 296 pp. 0-9723830-4-2 • $25.00

THE BARD ON THE BRAIN: Understanding the Mind Through the Art of Shakespeare and the Science of Brain Imaging

Paul Matthews, M.D., and Jeffrey McQuain, Ph.D. Foreword by Diane Ackerman

Explores the beauty and mystery of the human mind and the workings of the brain, following the path the Bard pointed out in 35 of the most famous speeches from his plays. 100 illustrations.

Cloth 248 pp. 0-9723830-2-6 • $35.00

STRIKING BACK AT STROKE: A Doctor-Patient Journal

Cleo Hutton and Louis R. Caplan, M.D.

A personal account with medical guidance from a leading neurologist for anyone enduring the changes that a stroke can bring to a life, a family, and a sense of self. 15 illustrations.

Cloth 240 pp. 0-9723830-1-8 • $27.00

UNDERSTANDING DEPRESSION: What We Know and What You Can Do About It

J. Raymond DePaulo Jr., M.D., and Leslie Alan Horvitz.

Foreword by Kay Redfield Jamison, Ph.D.

What depression is, who gets it and why, what happens in the brain, troubles that come with the illness, and the treatments that work.

Cloth 304 pp. 0-471-39552-8 • $24.95
Paper 296 pp. 0-471-43030-7 • $14.95

KEEP YOUR BRAIN YOUNG: The Complete Guide to Physical and Emotional Health and Longevity

Guy McKhann, M.D., and Marilyn Albert, Ph.D.

Every aspect of aging and the brain: changes in memory, nutrition, mood, sleep, and sex, as well as the later problems in alcohol use, vision, hearing, movement, and balance.

Cloth 304 pp. 0-471-40792-5 • $24.95
Paper 304 pp. 0-471-43028-5 • $15.95

THE END OF STRESS AS WE KNOW IT

Bruce McEwen, Ph.D., with Elizabeth Norton Lasley

Foreword by Robert Sapolsky

How brain and body work under stress and how it is possible to avoid its debilitating effects.

Cloth 239 pp. 0-309-07640-4 • $27.95
Paper 262 pp. 0-309-09121-7 • $19.95

IN SEARCH OF THE LOST CORD: Solving the Mystery of Spinal Cord Regeneration

Luba Vikhanski

The story of the scientists and science involved in the international scientific race to find ways to repair the damaged spinal cord and restore movement. 21 photos; 12 illustrations.

Cloth 269 pp. 0-309-07437-1 • $27.95

THE SECRET LIFE OF THE BRAIN

Richard Restak, M.D.

Foreword by David Grubin

Companion book to the PBS series of the same name, exploring recent discoveries about the brain from infancy through old age.

Cloth 201 pp. 0-309-07435-5 • $35.00

THE LONGEVITY STRATEGY: How to Live to 100 Using the Brain-Body Connection

David Mahoney and Richard Restak, M.D.

Foreword by William Safire

Advice on the brain and aging well.

Cloth 250 pp. 0-471-24867-3 • $22.95
Paper 272 pp. 0-471-32794-8 • $14.95

STATES OF MIND: New Discoveries about How Our Brains Make Us Who We Are

Roberta Conlan, Editor

Adapted from the Dana/Smithsonian Associates lecture series by eight of the country's top brain scientists, including the 2000 Nobel laureate in medicine, Eric Kandel.

Cloth 214 pp. 0-471-29963-4 • $24.95
Paper 224 pp. 0-471-39973-6 • $18.95

THE DANA FOUNDATION SERIES ON NEUROETHICS

DEFINING RIGHT AND WRONG IN BRAIN SCIENCE: Essential Readings in Neuroethics

Walter Glannon, Ph.D.

The fifth volume in The Dana Foundation Series on Neuroethics, this collection marks the five-year anniversary of the first meeting in the field of neuroethics, providing readers with the seminal writings on past, present, and future ethical issues facing neuroscience and society.

Cloth 350 pp. 978-1-932594-25-6 • $15.95

HARD SCIENCE, HARD CHOICES: Facts, Ethics, and Policies Guiding Brain Science Today

Sandra J. Ackerman, Editor

Top scholars and scientists discuss new and complex medical and social ethics brought about by advances in neuroscience. Based on an invitational meeting co-sponsored by the Library of Congress, the National Institutes of Health, the Columbia University Center for Bioethics, and the Dana Foundation.

Paper 152 pp. 1-932594-02-7 • $12.95
ISBN-13: 978-1-932594-02-7

NEUROSCIENCE AND THE LAW: Brain, Mind, and the Scales of Justice

Brent Garland, Editor. With commissioned papers by Michael S. Gazzaniga, Ph.D., and Megan S. Steven; Laurence R. Tancredi, M.D., J.D.; Henry T. Greely, J.D.; and Stephen J. Morse, J.D., Ph.D.

How discoveries in neuroscience influence criminal and civil justice, based on an invitational meeting of 26 top neuroscientists, legal scholars, attorneys, and state and federal judges convened by the Dana Foundation and the American Association for the Advancement of Science.

Paper 226 pp.1-932594-04-3 • $8.95

BEYOND THERAPY: Biotechnology and the Pursuit of Happiness. A Report of the President's Council on Bioethics

Special Foreword by Leon R. Kass, M.D., Chairman.
Introduction by William Safire

Can biotechnology satisfy human desires for better children, superior performance, ageless bodies, and happy souls? This report says these possibilities present us with profound ethical challenges and choices. Includes dissenting commentary by scientist members of the Council.

Paper 376 pp. 1-932594-05-1 • $10.95

NEUROETHICS: Mapping the Field. Conference Proceedings.

Steven J. Marcus, Editor

Proceedings of the landmark 2002 conference organized by Stanford University and the University of California, San Francisco, and sponsored by The Dana Foundation, at which more than 150 neuroscientists, bioethicists, psychiatrists and psychologists, philosophers, and professors of law and public policy debated the ethical implications of neuroscience research findings. 50 illustrations.

Paper 367 pp. 0-9723830-0-X • $10.95

IMMUNOLOGY

RESISTANCE: The Human Struggle Against Infection

Norbert Gualde, M.D., translated by Steven Rendall

Traces the histories of epidemics and the emergence or re-emergence of diseases, illustrating how new global strategies and research of the body's own weapons of immunity can work together to fight tomorrow's inevitable infectious outbreaks.

Cloth 219 pp. 1-932594-00-0 $25.00
ISBN-13: 978-1-932594-00-3

FATAL SEQUENCE: The Killer Within

Kevin J. Tracey, M.D.

An easily understood account of the spiral of sepsis, a sometimes fatal crisis that most often affects patients fighting off nonfatal illnesses or injury. Tracey puts the scientific and medical story of sepsis in the context of his battle to save a burned baby, a sensitive telling of cutting-edge science.

Cloth 225 pp. 1-932594-06-X • $23.95
Paper 225 pp. 1-932594-09-4 • $12.95

ARTS EDUCATION

A WELL-TEMPERED MIND: Using Music to Help Children Listen and Learn

Peter Perret and Janet Fox

Foreword by Maya Angelou

Five musicians enter elementary school classrooms, helping children learn about music and contributing to both higher enthusiasm and improved academic performance. This charming story gives us a taste of things to come in one of the newest areas of brain research: the effect of music on the brain. 12 illustrations.

Cloth 231 pp. 1-932594-03-5 • $22.95
Paper 231 pp. 1-932594-08-6 • $12.00

Free Educational Books

(Information about ordering and downloadable PDFs are available at *www.dana.org.*)

PARTNERING ARTS EDUCATION: A Working Model from ArtsConnection

This publication describes how classroom teachers and artists learned to form partnerships as they built successful residencies in schools. *Partnering Arts Education* provides insight and concrete steps in the ArtsConnection model. 55 pp.

ACTS OF ACHIEVEMENT: The Role of Performing Arts Centers in Education

Profiles of more than 60 programs, plus eight extended case studies, from urban and rural communities across the United States, illustrating different approaches to performing arts education programs in school settings. Black-and-white photos throughout. 164 pp.

PLANNING AN ARTS-CENTERED SCHOOL: A Handbook

A practical guide for those interested in creating, maintaining, or upgrading arts-centered schools. Includes curriculum and development, governance, funding, assessment, and community participation. Black-and-white photos throughout. 164 pp.

THE DANA SOURCEBOOK OF BRAIN SCIENCE: Resources for Teachers and Students, Fourth Edition

A basic introduction to brain science, its history, current understanding of the brain, new developments, and future directions. 16 color photos; 29 black-and-white photos; 26 black-and- white illustrations. 160 pp.

THE DANA SOURCEBOOK OF IMMUNOLOGY: Resources for Secondary and Post-Secondary Teachers and Students

An introduction to how the immune system protects us, what happens when it breaks down, the diseases that threaten it, and the unique relationship between the immune system and the brain. 5 color photos; 36 black-and-white photos; 11 black-and-white illustrations. 116 pp. ISSN: 1558-6758

PERIODICALS

Dana Press also offers several periodicals dealing with arts education, immunology, and brain science. These periodicals are available free to subscribers by mail. Please visit *www.dana.org*.